THIS BOOK BELONGS TO:
MICHAEL ROSS

Hardware Verification with SystemVerilog

An Object-Oriented Framework

Mike Mintz
Robert Ekendahl

Hardware Verification with SystemVerilog

An Object-Oriented Framework

Cover art from the original painting "Dimentia #10" by John E.
Bannon, johnebannon.com

 Springer

Mike Mintz
Harvard, MA
USA

Robert Ekendahl
Somerville, MA
USA

Library of Congress Control Number: 2007923923

ISBN 0-387-71738-2   e-ISBN 0-387-71740-4
ISBN 978-0-387-71738-8   e-ISBN 978-0-387-71740-1

Printed on acid-free paper.

9 8 7 6 5 4 3 2 1

springer.com

For Joan, Alan, and Brian.
Thanks again for your patience.
                    Mike

For Chantal.
Thanks again for your understanding,
love, and active support.

And to Newton—and now Darwin.
For many more missed walks.
                    Robert

Contents

**Part II:
An Open-Source Environment with SystemVerilog69**

Chapter 6: Truss: A Standard Verification Framework 89

Part III:
Using OOP for Verification
(Best Practices)

Preface

This is the second of our books designed to help the professional verifier manage complexity. This time, we have responded to a growing interest not only in object-oriented programming but also in SystemVerilog. The writing of this second handbook has been just another step in an ongoing masochistic endeavor to make your professional lives as painfree as possible.

The authors are not special people. We have worked in several companies, large and small, made mistakes, and generally muddled through our work. There are many people in the industry who are smarter than we are, and many coworkers who are more experienced. However, we have a strong desire to help.

We have been in the lab when we bring up the chips fresh from the fab, with customers and sales breathing down our necks. We've been through software bring-up and worked on drivers that had to work around bugs[1] in production chips.

What we feel makes us unique is our combined broad experience from both the software and hardware worlds. Mike has over 20 years of experience from the software world that he applies in this book to hardware verification. Robert has over 12 years of experience with hardware verification, with a focus on environments and methodology.

What we bring to the task of functional verification is over three decades of combined experience, from design, verification, software development, and management. It is our experiences that speak in this handbook. It is our desire that others might learn and benefit from these experiences.

We have had heated discussions over each line of code in this book and in our open-source libraries. We rarely agree at first, but by having to argue our cases we arrive at what we feel are smart, efficient, flexible, and simple solutions. Most of these we have "borrowed" from the software industry but have applied to the field of verification.

We believe that the verification industry can benefit from the lessons learned from the software domain. By using industry-standard languages, the verification domain can adapt techniques and code from over twenty calendar years

---

[1] Features.

of software effort, the scope of which is nothing short of stunning. Many brilliant people have paved the way in the software field. Although the field of verification is much younger, we could benefit greatly from listening, learning, and adapting mature programming techniques to the production of products of the highest quality.

So why do we provide open-source software at our website, www.trusster.com? Open-source software is a key to uniting and increasing the productivity of our industry. There is almost no successful closed-source ("hard macro") intellectual property (IP), for a good reason. Without the ability to look at the source and edit as necessary, the task is much more difficult and the chances for success are slim.

We hope that you enjoy this book—and better yet, find its principles increasingly useful in daily practice. We look forward to your comments. Please keep in touch with us at www.trusster.com.

Mike Mintz

Robert Ekendahl

Cambridge, Massachusetts, USA

March 2007

Acknowledgments

It takes a village to raise a child, and it takes a village to create a book. There is a core family, and a few relatives, and a whole lot of helpful neighbors and friends. Once again, the authors would like to bow humbly to our village—in particular, to the global verification village.

This, our second book, shares many of the same reviewers and adds some new ones. They provided great comments on almost every chapter, both detailed and "big picture," helping to improve many sections substantially.

Michael Meyer was once again our main technical editor, turning our gibberish into English and making clear where we were unclear. This book would not have been readable without him.

We are truly grateful for all the reviewers, their time, and their suggestions during both the early and near final stages of the book. In particular, we thank Ed Arthur, Oswaldo Cadenas, Jesse Craig, Simon Curry, Thomas Franco, John Hoglund, Mark Goodnature, Tom Jones, James Keithan, Ajeetha Kumari, David Long, Bryan Morris, Nancy Pratt, Joe Pizzi, Dave Rich, Henrik Scheuer, Chris Spear, Peter Teng, Thomas Tessier, Greg Tierney, Igor Tsapenko, Gerry Ventura, Stephanie Waters, and Andrew Zoneball.

We are also grateful for the support and encouragement of the producers of the HDL simulators. In particular, we thank the following simulator companies—Cadence, Mentor Graphics, and Synopsys—for providing licenses to their products, so we could confirm that the examples in this handbook work.

Introduction

Coding is a human endeavor. Forget that and all is lost.

Bjarne Stroustrup, father of C++

There are several books about hardware verification, so what makes this book different? Put simply, this book is meant to be useful in your day-to-day work—which is why we refer to it throughout as a handbook. The authors are like you, cube dwellers, with battle scars from developing chips. We must cope with impossible schedules, a shortage of people to do the work, and constantly mutating hardware specifications.

We subtitled this book *An Object-Oriented Framework* because a major theme of the book is how to use object-oriented programming (OOP) to do verification well. We focus on real-world examples, bloopers, and code snippets. Sure, we talk about programming theory, but the theme of this book is how to write simpler, adaptable, reusable code. We focus mainly on OOP techniques because we feel that this is the best way to manage the ever-increasing complexity of verification. We back this up with open-source Verification Intellectual Property (VIP), several complete test systems, and scripts to run them.

We cover the following topics:

- SystemVerilog as a verification language

- A tour of the features and real-world facts about SystemVerilog

- How to use OOP to build a flexible and adaptable verification system

- How to use specific OOP techniques to make verification code both simpler and more adaptable, with reference to actual situations (both good and bad) that the authors have encountered

- Useful SystemVerilog code, both as snippets, complete examples, and code libraries—all available as open source

This handbook is divided into four major sections:

- *Part I* provides an overview of OOP concepts, then walks through the transformation of a block-level view of a typical verification system into code and classes.

- *Part II* describes two free, open-source code libraries that can serve as a basis for a verification system—or as inspiration for your own environment. The first, called Teal, is a set of utility classes and functions. The second, called Truss, is a complete verification system framework. Both are available as open source and are available at www.trusster.com.

- *Part III* describes how to use OOP to make your team as productive as possible, how to communicate design intent better, and how to benefit from "lessons learned" in the software world.

- *Part IV* describes several complete real-world examples that illustrate the techniques described in the earlier parts of this book. In these examples we build complete verification environments with makefiles, scripts, and tests. These examples can serve as starting points for your own environment.

For the curious, each of the chapters in Part I and Part III ends with a section called "For Further Reading," which recommends relevant landmark papers and books from both the hardware and software domains.[1]

[1.] The references in these sections, though not academically rigorous, should be sufficient to help you find the most recent versions of these works on the Internet.

 • • • • • • •

Background

The silicon revolution[1] has made computers, cell phones, wireless networks, and portable MP3 players not only ubiquitous but in a constant state of evolution. However, the major impediment to introducing new hardware is no longer the hardware design phase itself, but the verification of it.

Costs of $1M or more and delays of three to six months for new hardware revisions of a large and complex application-specific integrated circuits (ASICs) are common, providing plenty of incentive to get it right the first time. Even with field-programmable gate arrays (FPGAs), upgrades are costly, and debugging an FPGA in the lab is very complex for all but the simplest designs.

For these reasons, functional verification has emerged as a team effort to ensure that a chip or system works as intended. However, functional verification means different things to different people. At the 30,000-foot level, we write specifications, make schedules, and write test plans. Mainly, though, we code. This handbook focuses on the coding part.

White papers are published almost daily to document some new verification technique. Most of you probably have several papers on your desk that you want to read. Well, now you can throw away those papers! This handbook compresses the last ten years of verification techniques into a few hundred pages. Of course, we don't actually cover that decade in detail (after all, this is not a history book), but we have picked the best techniques we found that actually worked, and reduced them to short paragraphs and examples.

Because of this compression, we cover a wide variety of topics. The handbook's sections range from talking about SystemVerilog, to introducing OOP, to using OOP at a fairly sophisticated level.

---

[1] Moore's law of 1965 is still largely relevant. See "Cramming more components onto integrated circuits," by Gordon Moore, *Electronics*, Volume 38, Number 8, April 19, 1965.

What is Functional Verification?

Asking "what is functional verification?" brings to mind the familiar poster, "A View of the World from Ninth Avenue,"[1] in which the streets of New York City are predominant and everything beyond is tiny and insignificant. Every one of us has a different perspective, all of which are, of course, "correct." Put simply, functional verification entails building and running software to make sure that a device under test (DUT, or in layman's terms, the chip) operates as intended—before it is mass-produced and shipped.

We perform a whole range of tasks where the end goal is to create a *high degree of confidence in the functionality of the chip*. Mostly we try to find errors of logic, by subjecting the chip to a wide variety of conditions, including error cases (where we validate graceful error handling and ensure that the chip at least does not "lock up"). We also make sure that the chip meets performance goals, and functions in uncommon combinations of parameters ("corner cases"), and confirm that the chip's features—such as the register, interrupt, and memory-map interfaces—work as specified.

As with the view of New York City, the perspectives of every company, indeed even of the design and test teams within a company, will naturally be slightly different. Nevertheless, as long as the chip works as a product, there are a number of ways to achieve success. That's why this handbook does not focus on what the specific tasks are; you know what you have to do. Rather, we focus on how you can write your code as effectively as possible, to alleviate the inevitable pain of verification.

---

[1] Saul Steinberg, cover of *The New Yorker*, March 29, 1976.

Why Focus on SystemVerilog?

A major development in the field of functional verification is the increasingly mainstream use of OOP techniques. Basically, those of us in the verification field need those techniques to handle increasingly complex tasks effectively. While most of the techniques presented in this handbook are adaptable to any number of languages such as Vera or C++, we focus on SystemVerilog—the marriage of the Verilog programming language with OOP.

At its core, OOP is designed to manage complexity. All other things being equal, simpler code is better. Because of the flexibility inherent in using OOP, we can write code that is simpler to use, and therefore more adaptable. In short, we can write reusable code that outlives its initial use.

This handbook is all about providing techniques, guidelines, and examples for using SystemVerilog in verification, allowing you to make more use of some "lessons learned" by software programmers. We distill the important bits of knowledge and techniques from the software world, and present them in the light of verification.

A Tour of the Handbook

The four parts of this handbook provide a variety of programming tips and techniques.

- *Part I* walks through the main concepts of OOP, introducing how to transform your high-level "whiteboard" idea for a verification system into separate roles and responsibilities. The goal is to build appropriately simple and adaptable verification systems.

- *Part II* uses these techniques and presents two open-source code libraries for verification, called Teal and Truss. Teal is a utility package that is lightweight and framework agnostic. Truss is a verification framework that encourages the use of the canonical form described in Part I. Both are used by several companies and run under most simulators.

- *Part III* introduces the OOP landscape in a fair amount of detail. OOP thinking, design, and coding are illustrated by means of code snippets representative of problems that verification engineers commonly have to solve.

- *Part IV* provides several complete examples of verification test systems, providing real-world examples and more details on how the OOP techniques discussed are actually used. Part IV is all about code. While a handbook may not be the best vehicle for describing code, it can be a good reference tool. We show a relatively simple example of how the verification of a single block of the ubiquitous UART[1] can be done. Then we show how this block-level environment can be expanded to a larger system.

The authors sincerely hope that, by reading this handbook, you will find useful ideas, techniques, and examples that you can use in your day-to-day verification coding efforts.

For Further Reading

- On the topic of coding well, *Writing Solid Code*, by Steve McGuire, is a good tour of the lessons Microsoft has learned.

- *Principles of Functional Verification*, by Andreas Meyer, provides an introduction to the broad topic of chip verification.

- *Writing Testbenches: Functional Verification of HDL Models, Second Edition*, by Janick Bergeron, gives another view of the process of functional verification.

[1.] Universal asynchronous receiver-transmitter.

Part I: SystemVerilog and Verification (The Why and How)

This part of the handbook explores the use of SystemVerilog for verification and then look at the benefits and drawbacks of using SystemVerilog. In the next chapter we take a brief tour of the features of SystemVerilog.

Next, we weave three different themes together: the evolution of programming in general, the creation of object-oriented programming (OOP) techniques, and the evolution of functional verification. The reason we chose to look at these three themes is to show why OOP exists and how it can be harnessed to benefit verification.

A major theme of this handbook is to build a verification system in layers. OOP techniques are well-suited to this approach. In the last chapter of this section, we'll look at a canonical verification system by using a standard approach to building verification components.

Why SystemVerilog?

> *If you want to do buzzword-oriented programming, you must use a strongly hyped language.*
>
> *Mike Johns*

We, in the functional verification trade, write code for a living. Well, we do that, and also puzzle over code that has been written and that has yet to be written. Because functional verification is a task that only gets more complex as designs become more complex, the language we work in determines how well we can cope with this increasing complexity.

The authors believe that SystemVerilog is an appropriate choice for functional verification, but as with any choice, there are trade-offs. This chapter discusses the advantages and disadvantages of using SystemVerilog for functional verification. We'll look at the following topics:

- An abbreviated comparison of the languages and libraries available for functional verification

- Why SystemVerilog is an appropriate choice for verification

■   The disadvantages of using SystemVerilog

Overview

Coding for functional verification can be separated into two parts. One is the generic programming part, and the other is the chip testing part. The generic part includes writing structures, functions, and interactions, using techniques such as OOP to manage complexity. The chip testing part includes connecting to the chip, running many threads, and managing random variables.

The generic programming part becomes more and more crucial as the complexity of the hardware to be tested grows. While the problem of connecting to a more complex chip tends to grow only linearly, the overall problem of dealing with this increased complexity grows exponentially.

The authors believe the generic part of programming is served reasonably by SystemVerilog. The language's features and expressive capabilities make it usable for functional verification. As will be discussed in detail in later sections, the downside is that the language is immature, and compliance from one simulator to the next is inconsistent.

While SystemVerilog might be a little rough around the edges, it is a good way for those who are mainly hardware oriented to learn OOP. As with Verilog, threading is built in, and connection to the chip is relatively well thought out. Realize though, that the actual percentage of code devoted to these tasks is small.

These tasks of HDL connection and parallel execution generally increase linearly with the complexity of the chip. In other words, there are more wires to connect, more independent threads to run, more variables to constrain, and so on.

By contrast, it is much more difficult to make the complexity of a chip increase only linearly. So, as a verification system gets bigger, things tend to get out of hand quickly. Our ability to understand a complex verification system is often more important than how we actually connect to the hardware description language (HDL) wires.

So this handbook concentrates on the "How to make the code reasonable" part of programming. Sure, our examples are multithreaded and use virtual interfaces,[1] but the bulk of this handbook is about how to write understandable code.

SystemVerilog as a Verification Language

Several attempts have been made to move verification away from HDLs, such as Verilog or VHDL.[2] An HDL does a good job of spanning design concepts (called the *register transfer level*, or RTL) down to a few primitives that are used in great numbers to implement a design (called the *gate level*). However, HDLs are not adept at "moving up" in abstraction level to handle modern programming techniques. HDLs are concerned with creating silicon, not with programming. Specifically, HDLs do not provide for object-oriented concepts.

SystemVerilog makes a step in this direction, and can be used to verify a chip. However, it is not clear that such a large span of concepts as SystemVerilog tries to cover can be integrated well into a single language. This handbook provides advice and examples that the authors believe will maximize the programming features of the language, while minimizing the "clunky" parts.

Not surprisingly, there are many choices and trade-offs when you choose a verification language. The table on the following page briefly lists the pros and cons of various languages suitable for verification.

---

[1] We talk about virtual interfaces in the next chapter, but for now just know that they are the way to connect HDL wires with testbench OOP code.

[2] VHSIC (Very High-Speed Integrated Circuit) HDL.

Language	Pros	Cons
Verilog, VHDL	Simple, no extra license required	No class concept, no separation of verification and chip concerns
Cadence Specman "e"	Rich feature set	Effectively proprietary, nonorthogonal language design
OpenVera	OOP—"like", better feature set than HDL	Effectively proprietary, interpreted, lacking full OOP support
SystemVerilog	IEEE standard, OOP features, one simulator does HDL and HVL, C interface	Covers all aspects from gates to OOP, implementation compliance is weak, language is large, yet lacking full OOP support
SystemC (C++)	Mature language, open source, most often does not need a simulator	Big footprint, focus is on modeling, heavy use of templating, coverage and constraint system dominates coding, long compile times, clumsy connection to HDL
Teal/Truss (C++ form)	Mature language, good use of C++, open source, few source files	Not a product, no inherent automatic garbage collection
Homegrown PLI/C	Free, well known	Not usually multithreaded, usually called from HDL as a utility function

As we stress repeatedly through this handbook, the team must decide what features of which languages to use, and how. This handbook will show how best to use SystemVerilog's OOP features.

Main Benefits of Using SystemVerilog

A major benefit of SystemVerilog is that it provides a relatively painless introduction to OOP, allowing you to use as little or much of OOP as you feel comfortable with. To this end, SystemVerilog allows the concept of "code interface" versus "implementation," allowing someone reusing code to concentrate on the features the code provides, not on how the code is actually implemented.

SystemVerilog is well-marketed, with several books and experts. (A quick web search for "SystemVerilog" yielded over 365,000 references.) The language is a good stepping stone from Verilog to OOP, reusing a fair amount of the Verilog syntax.

Furthermore, SystemVerilog vendors are developing useful debugging tools, and because SystemVerilog can coexist with Verilog and VHDL, existing HDL code can be integrated easily.

Many companies have behavioral c-models of their core algorithms. For models with a simple integral interface, the DPI[1] can be used to run the code in SystemVerilog. Note that the current compliance and feature set are spotty, so be prepared that you may have to rewrite the code in SystemVerilog.

SystemVerilog allows a clean separation between HDL and OOP concerns. As will be explained further in the next chapter, the use of the `virtual interface` feature, along with new keywords such as `class` and `local`, can be used to support the OOP concerns.

Drawbacks of Using SystemVerilog

While there are many benefits to using SystemVerilog, there are naturally drawbacks as in any language. One drawback is that, by itself, SystemVerilog is not a solution. Even with the open-source verification libraries of Teal and Truss, you have to write code in a new language.

---

[1] Direct Programming Interface—SystemVerilog's API for connecting to C, and by extension to C++.

Another drawback, ironically, is that SystemVerilog is a rich language—with the "dangerous" power that this implies. There are many features and even sublanguages. Figuring out which subset to use is a daunting task.

Consequently, it can take time to learn how to use SystemVerilog effectively, even with the help of good FAE[1] teams from EDA[2] companies giving presentations on the language and their design methodology. You will have to find your own techniques within SystemVerilog. This, by the way, is not necessarily a bad thing.

The purpose of this handbook is to lessen the effects of these drawbacks—by providing proven OOP techniques from the software world, and by illustrating, through real examples, how they are applicable to functional verification.

SystemVerilog Traps and Pitfalls

This section of the handbook will probably be the most controversial. We will talk about the current state of the SystemVerilog language.

We do not advocate using every feature in the language. Perhaps, over time, the benefits of the features will bear out. But because this language is immature, there are some areas where caution is advised.

SystemVerilog is not Verilog

Realize that SystemVerilog and Verilog are two separate languages. While there is movement within the language committees to join the two languages together, this will happen in 2008 at the earliest. Why does this affect you? SystemVerilog has, for the most part, Verilog behavior (and its warts), but there are differences.

For example, the SystemVerilog language reserves new keywords that are likely to make your Verilog code fail to compile. Fortunately, simulator vendors provide a way to tag files as either Verilog or SystemVerilog.

---

[1] Field applications engineer.
[2] Electronic design automation.

Errors and run-time crashes

When you code in a new language, there will be syntax and run-time errors. The majority of the time the compiler will be correct. However, remember that the language is young and the compliance is evolving, so do not spend a large amount of time debugging. Do not be shy about calling your local FAE. To be more clear, the authors and the FAEs are on a first-name basis.

Five languages in one!

As you start to learn SystemVerilog, it becomes clear that several languages are melded into one. SystemVerilog includes a synthesizable subset, an assertions language, a constraint language, a coverage language, and an OOP language. Whew!

Each of these sublanguages has its own syntax, semantics, and features, with a limited sharing of idioms. Because this handbook is focused on OOP for verification, we will discuss only the SystemVerilog OOP sublanguage.

With the exception of the synthesizable subset and OOP, these other features have not been proven universally necessary. They might work great for specific situations but not for most others. In the next sections, we present arguments why they may not withstand the test of time.

The assertions language

The authors have used assertions for years. Well, to be clear, we have used *nontemporal* assertions. These are simple boolean expressions that must be true, otherwise the simulation ends. The following is an example:

```
assert (request && grant);
```

This use is fairly straightforward. However, as soon as time is involved, the assertions can become quite complex, approaching the impenetrable.[1]

```
sequence qABC; a ##1 b ##[0:5] c; endsequence : qABC
property pEnded; not (qABC.ended); endproperty : pEnded
```

[1.] This example is from a forum on www.verificationguild.com.

```
first_match (qA23B).pEnded) |-> c;
```

The mental effort required to understand such constructs is large. The mental effort to write such constructs is an order of magnitude larger. This means that only a few engineers are able to create assertions. The authors have worked in languages where the complexity of the language created a "priesthood," where only the anointed could understand the actual meaning of the code. While this might create a sense of job security for the priests, it is never good for accuracy and efficiency, because it stops discussions about the code.

In addition, often the assertion-writing effort itself is equal to—or exceeds—the actual design-coding effort. While it's true that formal tools[1] can then be used, the effort required can be large compared to the payoff. This was tried in the software domain, complete with formal program proofs, but such proofs are no longer used.

Temporal assertions are complicated to write and understand. Make sure that the HDL complexity requires their use.

The constraint language

In verification, randomization is essential. Unfortunately, it can be difficult to control the parameters that manage the randomization. (This topic is discussed in detail in later chapters.) It certainly is not clear that we, as an industry, understand enough about managing randomization to have a "best" solution. The random-number management solution used in SystemVerilog includes a *constraint language*. It is unclear to the authors that this is a benefit. Sure, at some level we have to constrain random numbers to a range (or disjoint ranges), and possibly skew the distribution so that it is nonuniform. However, adding a declarative sublanguage within a procedural verification language is not an obvious win. The declarative language may look deceptively procedural. In addition to requiring the verifier to learn an HVL, the application of hierarchical and overlapping constraints is not intuitive.

---

[1] Yes, commercial assertion libraries for standard protocols—when available— can sometimes be useful, but beware: writing your own can be tricky!

For example, in one company we used the recommended method of extending a class to add constraints. This is "obvious" in theory, but in a real system one often cannot find, or keep in mind, all the classes and their subclasses. We kept adding constraints that conflicted at run time, and other testers added constraints to a class that many people were already using—even though the added constraints were applicable to only a single test. Finally, we decided that all constraints were to be local to a class, and not in the inherited classes.[1]

There are two techniques the authors have used successfully to perform constraints. One technique uses procedural code to set up min/max variables for constraining the random variable, and the other uses a forward declaration on a constraint. The first technique will be used in the examples, and so is not discussed further here. For the forward declaration technique, we declare a constraint test, without a body, in every class that has random behavior.

```
class ethernet_packet;
///method and data declarations
int packet_size;
constraint test; //no implementation in this class
endclass
class usb_generator;
int device_id;
constraint test; //no implementation in this class
endclass
```

Then in the actual test case, we implement the specific class's test constraint that we need.

```
//in test_<some test name>.sv we now define constraints
constraint ethernet_packet::test { packet_size == 218;}
constraint usb_generator::test {device_id == 4;}
//the rest of the test code
```

This allows each test case to have "knobs," to control the code as appropriate.

---

[1] Don't worry if these terms are a bit confusing in this paragraph. They will be explained in the next chapter.

Use constraints sparingly, either as a min/max bounds or as an unimplemented constraint that a test may use.

The coverage language

In addition to using constraints to guide the randomization, SystemVerilog adds a *coverage* sublanguage. While coverage is a good idea in theory and is a well-marketed concept, the authors are not certain that the industry has a clear need for it as implemented. It is a relatively simple matter to collect data, but many questions remain:

- Do you keep the time at which the coverage event occurred?

- How do you fold a large coverage range (such as an integer or a real) into coverage bins?

- What is the relationship between the covered events and the constraints that control the randomization?

These points show the difficulty in using coverage. They are inherent issues with functional coverage, as contrasted with line, toggle or expression coverage. Since humans define what the function of a chip is, humans need to define the coverage of these functions. In other words, it can never be simpler than defining what the chip does, which is not an easy task.

There is one more question:

- Will your company delay the chip tape-out or FPGA[1] delivery if coverage goals are not met?

This last question is critical. Be honest in your assessment. After many years of working on chip projects, it's our honest assessment that most companies would be fiscally delinquent if a product were delayed because of the possibility of some bugs.[2] The fact is that a company needs revenue, and the chip should have been tested adequately for basic market features *at least*—assuming the verification team is reasonably competent. If the team isn't, your company has more pressing problems to deal with.

One final point: It's common for software drivers to have to work around major deficiencies in a chip, as well to work around minor deficiencies

---

[1.] Field-programmable gate array.

[2.] Now of course, there are exceptions. The medical and space industry come to mind.

in many chips. If the chip runs and even performs a subset of the features adequately, your company will sell it and make revenue.

So why use coverage at all? Coverage is good for your configuration parameters. There are modes, such as `baud_rate`, `data_width`, and so on, that are set once and then used throughout the test run. By looking at coverage data, you can see that your basic data-flow tests are properly walking the configuration space of the chip.

> *Use coverage in SystemVerilog as part of your basic data-flow tests, but be careful: This coverage does not necessarily increase the productivity of your team. Write directed tests (without coverage) for specific cases and data-flow patterns.*

SystemVerilog features not discussed

SystemVerilog has many features, some of which are essentially vendor-specific. Other features are just not universally implemented or well thought out. In this section we'll enumerate some of these features.

Within SystemVerilog, templating in the OOP sublanguage is like using parameters in HDL. While an interesting feature, the vendor support for templating is weak and real-world proofs are even weaker. Templating makes SystemVerilog OOP code more complex, and does not map to C++ templating, which is well-proven.

The bind construct is a fairly loose part of the specification at present. It is also primarily used to connect SystemVerilog assertions to the synthesizable subset, so it is not a focus for verification coders. It is also a declarative construct, making it inappropriate for run-time configurations.

There are many more minor features of SystemVerilog, such as wild equality, the `ref` concept, clocking blocks on interfaces, and new datatypes such as `byte`, `shortreal`, `int union`, `enum`, and `string`. Searching through the SystemVerilog specification for terms such as "SystemVerilog adds" and "SystemVerilog introduces" will produce a fairly complete list. Because this handbook is concerned primarily with OOP and SystemVerilog, we will not discuss these features further here.

Summary

This chapter made the case for using SystemVerilog as a verification language. We took a quick look at some other options and then enumerated why SystemVerilog was appropriate.

The main point of this chapter is that SystemVerilog is a relatively easy path from Verilog to OOP.

Because SystemVerilog is a new and evolving language, we spent a fair amount of time presenting notes of caution. We also took note of specialized and new features that are not in the mainstream of OOP.

For Further Reading

- *Software Engineering: A Practitioner's Approach*, by Roger S. Pressman, has a great section on the evolution of programming. This handbook also has references to landmark papers and books.

- The SystemC and Testbuilder manuals have discussions on why C++ is good for verification. SystemC information can be found at www.systemc.org, and Testbuilder information can be found at www.testbuilder.net.

- Teal and Truss were initially documented in the authors' other book, *"Hardware Verification with C++:A Practitioner's Approach*. The current version of the source code for C++ and SystemVerilog is available on www.trusster.com.

- There are several standards for verification and simulation, such as 1800 for SystemVerilog, IEEE 1364-1999 (for VHDL), IEEE 1995-2001 (for Verilog), IEEE 1076, and IEEE 1647 (for the IEEE version of Cadence Specman "e"). The website www.openvera.org provides the OpenVera specification.

- There are a growing number of books devoted to coding in SystemVerilog. One book that the authors have used is *SystemVerilog for Verification: A Guide to Learning the Testbench Language Features*, by Chris Spear. It is a good look at most of SystemVerilog's features. (Note that this book is specific to Synopsys, so *caveat emptor*.)

- If you want to learn more about SystemVerilog assertions, consider the *SystemVerilog Assertions Handbook* by Ben Cohen, Srinivasan Venkataramanan, and Ajeetha Kumari.

- If you want a detailed look at the evolution of the SystemVerilog language, sign up for the *SystemVerilog Testbench Extension Committee* mailing list, at *http://eda.org/sv-ec*.

- Stuart Sutherland has a great paper (from SNUG[1] Boston 2006) titled "Standard Gotchas: Subtleties in Verilog and SystemVerilog That Every Engineer Should Know," available at *http://www.sutherland.com/papers.html*.

---

[1] Synopsis Users Group.

OOP and SystemVerilog

Progress has not followed a straight ascending line, but a spiral with rhythms of progress and retrogression, of evolution and dissolution.

Johann Wolfgang von Goethe

The idea of progress in the art and science of verification seems simple enough —until you look at how progress is made. It is rarely a single person, technique, or language that moves us forward to simpler code, while handling ever more-complex chips. Rather, it is a swinging, jumping roller-coaster that we are on. OOP is just another of those twists and turns along the ride of progress.

This chapter looks at why and how object-oriented programming was developed, and reflects on why OOP is the right choice for managing the increasing complexity of verification. It then shows how OOP is expressed in SystemVerilog. The OOP techniques shown in this chapter are used throughout the remainder of this handbook.

Overview

OOP is a programming technique that is often touted as a cure-all for verification. While it is true that OOP is an essential tool in a programmer's toolbox, it is by no means the most important one. One's experience, intelligence, and team environment are far more important to the success of verification than any language feature or technique. That said, OOP is a useful tool for communicating and enforcing design intent for large projects and teams, in addition to being a good way to build adaptable, maintainable, and reusable code.

This handbook is intended for those having at least some familiarity with the concept of OOP. Many verification engineers already have some experience with OOP through languages such as C++, Vera, Specman "e," or SystemC.

The first part of this chapter looks at the history of OOP and why it is well-suited to functional verification. The second part shows how SystemVerilog expresses the most common elements of OOP.

For readers with limited experience in OOP, there are a few suggestions at the end of this chapter. If you have at least some experience with OOP, or if some time has passed since you used it last, then *don't worry!*

Some of the aspects presented in this and subsequent chapters might seem confusing at first, but Part II of this handbook shows a complete working verification environment. It is the authors' hope and intent that you will "copy and paste" from this environment as well as from the examples provided.[1] This handbook is designed to give you a jump start on using SystemVerilog without having to design every class from scratch.

The "basic" OOP techniques expressed in this chapter are important, and form the basis of the fancier techniques in Part III of the handbook.

[1.] Code is freely available at www.trusster.com.

The Evolution of OOP and SystemVerilog

OOP techniques have been proven to help large programming teams handle code complexity. One key to coping with such complexity is the ability to express the *intent* of the code, thus allowing individual programmers to develop their part of the code more effectively. This understanding of intent allows programmers to build upon already working code, and to understand the overall structure more easily.

Assembly programming: The early days

Programming has changed a lot over the years. It started with the use of assembly language[1] as a way to express a "simple" shorthand notation for the underlying machine language. This simple abstraction allowed programmers to focus on the problem at hand, instead of on the menial and error-prone task of writing each instruction as a hexadecimal or octal integer. Simply put, abstraction allowed an individual programmer to become more productive.

Here is an example of some assembly language:

```
            MOV.W     R3,  #100
            MOV.L     R1,  #7865DB
    loop:   ADDQ.W    R1,  #4
            TST.W     R1,  R2
            BNZ       loop
```

Procedural languages: The next big step

With the increase in complexity of the problems programmers were asked to handle, procedural languages such as FORTRAN,[2] C, and Pascal were developed. These procedural languages became very popular and allowed individual programmers to become highly productive.

Here is an example of FORTRAN,[3] a common procedural language:

---

[1] The first assembly language was created by Grace Hopper in 1948.

[2] For FORmula TRANslator, created by John W. Backus in 1952.

[3] Okay, you got us—this is actually FORTRAN 77, the "new" FORTRAN (ANSI X3.9, 1978).

```
          DO 3, LOOP = 1, 10
          READ *, MGRADE, AVERAGE
          IF (.NOT. (AVERAGE .GT. 6.0 E -1)) THEN
            PRINT *, 'Failing average of ', AVERAGE
            STOP
          ELSE
            PRINT *, 'Passing average of', AVERAGE
            AVERAGE = (MGRADE / 1 E 2) + AVERAGE
          END IF
        3 CONTINUE
```

Interestingly, as the size of the programs grew, the focus of programming switched from the productivity of the individual to the productivity of the larger team. It was found that procedural languages were not well-suited to large programming efforts, because communicating the intent of the code was difficult. OOP, with its ability to build classes upon classes and define interfaces, proved an effective response to this problem.

OOP: Inheritance for functionality

By necessity, OOP developed in stages. The first stage focused on what is often called *data hiding* or *data abstraction*. This is a way to organize large amounts of code into more manageable pieces. With large amounts of procedural code, it became very complicated to keep track of all structures and the procedures that could operate on those structures. It was also hard to expand, in an organized way, upon existing code without directly editing the code—a process that, as we all know, is error prone.

To address these problems, a language called Simula was developed in 1967. This language is recognized as the first language to introduce object-oriented concepts.

SystemVerilog has this lineage, with ways to organize data structures and the functions that operate on those structures. This organizational concept is called a *class* (loosely based on Simula's class). The tasks and functions, now scoped within a class, are called *methods*. In addition, SystemVerilog included ways for one class to expand upon another through *inheritance* (also from Simula).

*The very essence of OOP is the ability to specify similarities
and differences in code constructs relatively easily.*

Classes allowed for the grouping of code with data, while inheritance allowed a way to express increasingly intricate functionality through the reuse of smaller working modules. This technique is often called *inheritance for functionality.* (Later in this chapter, we'll show how System-Verilog expresses both of these features—grouping into classes and reuse through inheritance—in more detail.) This new approach was sort of like the Industrial Revolution of the programming world, increasing team productivity by an order of magnitude.

Classes helped improve the productivity of programming teams by organizing the code in layers—with one layer inheriting from, and enhancing upon, a lower layer. This meant that the code could now be "reasoned about." With "reasonable" code, changes and bug fixes could be made only to the appropriate lines, without the changes echoing, or propagating undesirably, throughout all of the code.

Furthermore, as code was structured into layers through hierarchy trees, several patterns became visible. For example, it became clear that certain layers were not involved with manipulating the data (in the classes) directly, but rather with ordering, structuring, and tracking events.

These framework layers became more and more important to understanding the system. To get a large program to be "reasonable," more and more standard infrastructure was needed. These framework layers had no "interest" in how the actual data were manipulated; rather, the important feature was that now the data could be assumed to be manipulated in predefined ways.

As an example, as long as each class in a particular framework layer had a `start()` or a `randomization()` function, working with classes of that type was reasonable. As these framework layers were written, it became clear that they could be generalized as long as each class followed the rules for that type of "component."

OOP: Inheritance for interface

So how to get a class to "follow the rules" of a framework component? What is needed is a language-enforced way to express the rules that a class had to follow in order to "fit in." The solution, known as *virtualization*, is included in SystemVerilog. With virtualization one could define classes called *virtual base classes*; these simply express the *code interface* to which a component must conform, in order to fit into the larger system.

Each developer of the actual classes that fit in a particular structure would then inherit from this virtual base class, and implement the details for how a particular function should be implemented for the problem at hand. This technique of defining the code interface through virtualization, often called *inheritance for interface*, is frequently used in OOP-based projects.

The clever thing is that now one could write the code for the framework layer using virtual base classes. This not only allowed the framework to be implemented concurrently with the data-based classes, but it also allowed the framework layer to be developed in a much more generic way. This virtualization of base classes has proven to be a powerful technique for creating and maintaining large and complex systems.

A word or two about "interface"

It is unfortunate that SystemVerilog has a keyword called `interface`. This is because "interface" is a common term in OOP for expressing the class items (data and methods) with which a user is concerned. This is also called the `public` part of a class, but interface is a more wide-spread term. We will talk more about the SystemVerilog `interface` at the end of this chapter.

So, in this handbook, we will mostly use the phrase "code interface" to refer to the public code of a class, and use `interface` (in that weird code font) to indicate the SystemVerilog keyword. When we feel the context is sufficient, we may omit the distinction here and there.

The Evolution of Functional Verification

Verification through inspection

There are similarities with the development of OOP and that of functional verification, and while hardware verification is a younger field than software programming, it has (not surprisingly) followed a similar path.

As readers of this handbook surely know, functional verification has come a long way from its recent humble beginning as a (mostly manual) process of verifying simulation waveforms. From there, it evolved into "golden" files; a current simulation run was compared to a known-to-be-good result file—the golden file. For this technique to work it required fixed stimuli, often provided in simple text format. Golden files were an acceptable technique for small designs, where the complete design could be tested exhaustively through a few simulation runs.

Verification through randomness

The simple technique of using golden files became impossible to use as the size of the hardware being tested grew both in size and complexity, so other techniques were needed. For larger projects it was no longer possible to test the "state space" of a chip completely. To do so would require an unobtainable amount of computer time, even on the fastest machines. To address the reality that the chips being developed could no longer be tested exhaustively, random testing was introduced. Using randomness in tests changes the input stimuli every time a test is run. The goal is to cover as much of the state space as possible through ongoing regression runs.

Unfortunately, several problems were found in using randomness with current hardware description languages (such as Verilog or VHDL). To begin with, the result checking became more complex as golden files could no longer be used (because the input stimuli changed for each run). This meant that verification models that could predict the outcome from any set of input stimuli were needed. Writing these models has become a major task for the verification projects of today.

However, this technique also posed other problems. It was discovered that using randomness was a tricky thing. If you use random stimuli and your testing fails because of a hardware bug, then you later would want to rerun the exact same sequence of stimuli to verify that the bug has been solved. This is more easily said than done.

You can record all the stimuli that generated the test run, then use some mechanism to replay the stimuli later; alternatively, you can track the "seed" from which the (pseudo) random generator starts and then pass that number into your next simulation run.

Both techniques can be problematic, because storing all the generated stimuli requires a lot of disk space and directory infrastructure, and because controlling randomness through a seed requires good control over your "random" generator.

The current most common solution to this problem is to control and store the "random" seed, then use it to replay a given stimuli sequence over and over.

The emergence of
hardware verification languages

We can see that controlling the generation of random stimuli requires many things. We need verification models that can predict results from any given set of stimuli. We also need control over how the random generator works, to be able to replay a given stimuli sequence. It was found that using HDL languages, such as Verilog and VHDL, was difficult with respect both to writing high-level models quickly and controlling randomness. In Verilog, for example, it was not obvious how to control the random seed back in 1987.

As a result, people started looking at other languages for verification. The natural first step was connecting C to Verilog, but soon languages such as "e" and Vera were introduced. These languages made it easier to do random testing, in turn making it possible to test much larger chips.

OOP: A current trend in verification

The problems we are facing today in verification are similar to the problems software faced when OOP was adopted. We now have to deal with very large amounts of code and multitudes of modules, all of which must be compiled, instantiated, controlled, randomized, and run. This is not an easy task, and we spend more and more time solving these basic framework problems. Specman "e" and Vera were early and proprietary entries in OOP-enabled hardware verification languages (HVLs). SystemVerilog is the latest entry, and promises a multivendor descendency.

It seems clear that adopting OOP techniques should help make these problems more manageable. Unfortunately, there are still not enough people in the field of verification who have sufficient experience and understanding of how to develop an appropriate OOP infrastructure.

Engineers in our field are just starting to adopt OOP techniques. The main reason for this book is to show verification techniques through "OOP glasses."

OOP: A possible next step

The field of verification is young; not long ago we were staring at waveforms on a screen. By using modern verification languages we have developed the field into something better. However, today we are facing even harder problems, one of which is the issue of the *framework*. To do a job that is increasingly complex, we need a framework for how our verification environment is interconnected. This is no longer an easy thing to achieve. In this handbook we show many techniques for how to manage this and other problems. We also introduce an open-source verification framework, called *Truss,* which collects our best experience in OOP into a working environment.

It is our belief that if enough people adopt a powerful open-source infrastructure, many great innovations will result. The problem we face today cannot be solved by the features of individual languages alone; rather, we need an agreed-upon framework. Even if this framework were modified by each team, it still provides the opportunity for best practices to evolve. This handbook, and the associated open-source code, is our attempt to start the discussion.

However, we are getting ahead of ourselves, so before we dive into the practical problem of verification, let's look at how SystemVerilog expresses OOP techniques.

OOP and SystemVerilog

This section shows how SystemVerilog expresses the OOP concepts described above. It describes some of the techniques we use to build a successful verification environment in later chapters. For engineers experienced with other OOP languages such as C++, Vera or "e," this chapter can serve as a way to map concepts from one language to another.

Data abstraction through classes

Using classes to express data abstraction is an important technique in building large verification systems. Data abstraction, by grouping the data and the operations together, allows engineers to reason about the code.

We will look at a direct memory access (DMA) descriptor class to show how a class can be constructed, then evolved by means of inheritance.

A DMA descriptor example

DMA is a common hardware feature for transferring data from one memory location to another without putting a load on the CPU. In this example, we verify a DMA chip that accepts DMA descriptors, puts them into an on-chip memory array, and then executes them. Each descriptor has a source and destination memory address, as well as the number of bytes (called "length") to transfer.

In the verification environment, a DMA descriptor could be represented by a small class. The DMA generator is then responsible for building, or *instantiating*, DMA descriptors and "pushing" them to the chip and to the checker.

The following code describes the DMA descriptor class:

```
class descriptor;
   //Constructor
   extern function new (int src, int dest,
                        int length, int status);
   //Code (or Public) Interface:
   extern virtual function void print();
   int source_address_; //public as an example!
   extern virtual function bit equal(descriptor d);
   //Implementation (or local and protected) interface:
   extern local virtual function int unique_id ();
   protected int destination_address_;
   protected int length_;
   protected int status_;
   protected int verif_id_;
endclass
```

The descriptor class is divided into the code (or public) interface and the implementation (or local and protected) interface, as shown by the *access control* level on each line. The next section will explain what access control labels are for.

Access control

The keywords `protected` and `local`, as used in the `descriptor` class, are SystemVerilog *access control* labels. They indicate how methods and variables following the statements can be used. The absence of a label indicates that the methods and variables that follow are publicly accessible by any code that has access to an instance of the class.[1] A `local` label indicates that only the code inside the class itself can access the variable or method.

The keyword `protected` indicates a private variable or method that can be modified through inheritance. Public, local, and protected can be used to express and enforce the intent of the class quite clearly.

---

[1] This is an unfortunate default, as a vast majority of classes will have far fewer public methods and variables compared to the local and protected code used to implement the class. Be prepared to type "local" and "protected" a lot.

Access control is needed to help separate the user or code interface from both the internal methods and the data needed to implement the class. Consequently, the *public* section of a class declaration is the "code interface." These are the interesting methods and variables to look at when you want to *use* a new class. When you *implement* a class, on the other hand, you also need a space to store the "state" of your class between method calls. This is done in local or protected scope.

Implementing a class is similar to implementing a state machine, where each method call changes the state of the state machine (that is, modifies the data members of the class). This "change of state" must be recorded somehow. Variables for tracking the state as well as intermediate methods should be put in the *local* scope, not only to indicate to users that they shouldn't focus on these methods and variables, but also to protect these variables from accidentally being modified. When a class is instantiated, only the public methods and calls can be accessed. Trying to access local scope results in an error during compilation. This is an example of how language enforces the "intent" of the class.

Enforcing intent can (and should) go beyond protecting state variables. For example, instead of printing an error message during run-time, when the code calls internal implementation-detail methods, one should declare those methods to be in local scope, so that a compile error occurs instead.[1]

Constructors

When a class is instantiated, the special function `new()` is called. This special function is the *constructor*. A constructor is used to initialize member variables, reserve memory, and initialize the class.

So how do you actually create an instance of a class? Consider our descriptor class for a moment. The class could be instantiated as follows:

---

[1] Note that SystemVerilog does access checking first, then resolution checking. This is unfortunate, as it means the code can behave differently when the access control is changed. This problem is discussed in detail in the book *The Design and Evolution of C++*.

```
descriptor descriptor1 = new (source_addr,
                              destination_addr,
                              source, length);
descriptor descriptor2;
descriptor descriptor2 = descriptor1;//point to the same object!
descriptor descriptor3;   //Careful: null pointer!
```

The first line declares `descriptor1`, and calls the constructor method as declared above, passing in variables as necessary. Next, `descriptor2` is created and then is just assigned the pointer to `descriptor1`, meaning that `descriptor2` is exactly the same as `descriptor1`. Note that the next descriptor, `descriptor3`, is not initialized at all, so SystemVerilog assigns a special keyword, `null`, to the variable, because constructors are not automatically called in SystemVerilog.

Each line is valid SystemVerilog, but this might not be what you intended.

> *Be aware that there are different ways of initializing an instance pointer.*

When writing a class, try to express the intent of the class so that an unintended use of your class generates a compile error. Though annoying, compile errors are much easier to understand than run-time errors. Similarly, when you get an unexpected compile error, don't see it as an annoyance, but rather realize that the person who wrote the class may be trying to tell you something.

Member methods and variables

In the `descriptor` class example, a few member variables and member methods are declared. The member variables are simply integers for the fields of the DMA transaction. The `print()` method will simply print all the current fields of the projects. Member variables and methods can be accessed like this:

```
descriptor d1 = new ('h68000, 'h20563, 39, 0);
d1.source_address_ = 'h586;
d1.print ();
```

Inheritance for functionality

By using class inheritance, you can create larger and larger functional blocks, building upon existing functionality. By inheriting from another class, you are saying, "I want to start from the functionality of an existing class and expand upon or change it with the features I define in my new class."

Consider our DMA project again. In the first generation of the product (as described above), the chip would simply store DMA descriptors in an array and signal when the array was full. For the second generation, this has been improved and the chip now implements a linked list, storing each descriptor in off-chip memory.

To enable this functionality, a hardware pointer field must be added for each descriptor. As a technique, the pointer can be set to 0 to stop the chip from processing, or it can be set to the first descriptor to implement a ring.

Instead of copying and editing the descriptor class, we can simply inherit from and expand upon the base descriptor class. This is called *inheritance for functionality*.

To create our new, fancy `linked_list_descriptor` class, we could declare it like this:

```
//in linked_list_descriptor.svh
'include "descriptor.svh"
class linked_list_descriptor extends descriptor;
    extern function new (uint32 src, uint32 dest,
                            uint32 length, uint32 next);
    extern virtual function void print();
    local uint32 next_descriptor_; //Pointer to DMA memory
endclass
```

The `extends` keyword from the first line of the class states that the `linked_list_descriptor` class inherits from the descriptor class. Now the `linked_list_descriptor` class has all the functionality of the original descriptor class, and adds the `next_descriptor` variable.

Note that in an extended class, you must manually call the base class constructor:

```
//in linked_list_descriptor.sv
function linked_list_descriptor::new
     (uint32 src, uint32 dest, uint32 length, uint32 next);
     super.new (src, dest, length, 0);
     next_descriptor_ = next;
endfunction
```

Inheritance for code interface

As we have seen, *inheritance for code interface* means using a base class, with virtual methods, to describe the class framework. This base class identifies some or all of the methods as *virtual*; classes extending from the base class can, and sometime must, implement those virtual methods.

With this technique, standard code interfaces for similar, but different, components can be used. This is very useful in creating a verification framework, because in a large verification environment you must keep track of a large number of components—for example, verification components [such as bus functional models (BFMs), generators, checkers, and monitors], and test components. By defining a code interface to which each type of component must conform, the ability to reason about the environment increases.

It should be noted that defining appropriate virtual base classes is not easy. Overly complicated or overly generic base classes tend to make the problem of verification more confusing instead of less. In Part III of this handbook we'll talk about the trade-offs.

As an example of *inheritance for code interface*, let's consider building a virtual base class for a BFM for a verification project. The team could decide that all verification components need certain *phases* (expressed as method calls), including `do_randomize()`, `out_of_reset()`, `start()`, and `final_report()`.

These methods ensure that a verification component is randomized, has time to program its part of the chip, starts up any threads needed to run, and has a way to print its status once the simulation is done. This can be done by creating a *virtual base class* from which all actual drivers inherit.

A virtual[1] base class in our example could look something like the following:

```
virtual class verification_base;
  virtual task out_of_reset ();/*do nothing */
    endtask
  virtual task do_randomize (); /*do nothing */
    endtask
  'PURE[2] virtual task start (); //NO implementation
  'PURE virtual task final_report ();
  //NO implementation
endclass
```

What makes the class virtual is the fact that the class starts with the `virtual` keyword, and at least one member method is declared by means of the keywords `pure` and `virtual`. In this example, the class declares how the verification system framework expects any verification components to behave, by enforcing that all verification component have at least these methods.

There are two types of virtual functions: *virtual* and *pure virtual*. In the example above, the first two methods are virtual, the last two are *pure* virtual. Virtual functions have a "default" implementation (in our case they do nothing), while pure virtual functions have no implementation and are indicated by "`pure`." A pure virtual function is one that a derived class is obliged to implement. For virtual methods, the original method is used if no same-named method is declared.

Consider an Ethernet driver, which is inherited from the class `verification_base`.

```
class ethernet_driver extends verification_base;
  task out_of_reset();
    set_up_dut();
  endtask
  task start();
  task final_report();
endclass
```

[1.] In SystemVerilog, the keywords `virtual class` mean "abstract base class" in an OOP sense.

[2.] Because `pure` is not yet a keyword in SystemVerilog, Truss uses the macro `` `PURE ``, which will work now and in the future.

　• • • • • • •　　Hardware Verification with SystemVerilog

In our Ethernet driver, `do_randomize()` is not declared, so the default method specified in `verification_base` is used.

This technique of using inheritance for code interface is very important for creating a flexible, yet reasonable, verification structure. In our verification domain, many objects must be initialized through many phases, synchronized, and run. This is not easy for anything but the smallest projects. However, by using inheritance for code interface and virtual methods, one can create a powerful and flexible verification environment.

What's a header file?

Throughout this handbook we refer to header files and source files. This is a widely used convention where the class declaration has all methods declared with the keyword `extern`, which means there is no body of code statements to the method. The file that has the class declaration like this is called a header file and usually has the extension `.svh`. There is a corresponding source file, which contains all the method definitions and usually has the extension `.sv`. Note the following:

```
//in bfm.svh...
class a_bfm;
  extern function new (string name,
                       virtual interface_wires pins);
  extern task write (int address, int data);
  extern task read (int address, out int data);
endclass

//in bfm.sv
a_bfm::new (string name, virtual interface_wires pins);
  //cache string and the interface
endfunction
task a_bfm::write (int address, int data);
  //perform a write to the bus
endtask
task a_bfm::read (int address, int data);
  //perform a read from the bus
endtask
```

Why go to the trouble of creating both a header file and an implementation file? To make it easier for other users of your code to reason about the class. They should spend their effort learning what the code interface is, from the header file, and not how you actually implemented the class in the source file.

Use header files (`*.svh`) and implementation files (`*.sv`) to improve the readability of your code.

Packages

It happens in every large verification project. You try to link all compiled files together and run into conflicting variable names; it seems that there is always more than one module called "generator" or "driver." It's frustrating, because now you have to go back and rename the conflicting classes and files. Furthermore, the "new rules to follow" probably becomes "you must insert interface name before variable name," so you end up with `uart_generator` and `ethernet_driver`.

But how do you deal with code from Intellectual Property (IP) vendors? How do you know what names *they* use?

SystemVerilog has a solution to this common problem: the use of *packages*. A package is the placement of related classes and global functions in a logical group. For example, if you are testing an Ethernet protocol, all your classes and components might go into the `ethernet` package. If you are testing a UART interface, consider using the `uart` package, and so on.

A package is simply declared as follows:

```
package pci_x;
  class master;

    . . .

  endclass
endpackage
```

Any class or variable wrapped inside the `package/endpackage` is now in the `pci_x` package. When you later want to instantiate a `pci_x` master, you simply declare what package you are using and what module you want.

Note, for example, the following:

```
pci_x::master my_master = new ("master pci_x");
```

The `pci_x::` indicates that you want to instantiate the `master` class from the `pci_x` package. If you are using a lot of components from a certain package, you can declare that you want to have access to that package throughout your file[1] by means of the keyword `import`, as follows:

```
import pci_x::*;
```

From that point on, you have access to all components in the `pci_x` package.

However, be careful about putting an `import` clause in a header file (a file with an `.svh` extension). *This is almost always a mistake.* The reason is that the `import` clause has now been added to every file that directly or indirectly includes this header file. So, the fact that some code was in a package is now lost to code that includes this header file. The authors have been on projects where using `import` in header files caused the very name collisions that packages were designed to avoid.

There are a number of caveats about working with packages. One is that there can be only a single `package/endpackage` declaration. This means that you often end up including all the header files in a meta-header file.

```
package pci_x;
  `include "pcix_master.svh"
  `include "pcix_slave.svh"
  `include "pcix_generator.svh"
  `include "pcix_monitor.svh"
  `include "pcix_driver.svh"
endpackage
```

Also, while you can use the `extern` keyword on class methods included in a package, their definition must also be within the `package/endpackage`. This is clumsy and can cause difficulties because of `` `include `` file dependency order. As a convention, the authors put the `` `include `` of the implementation file as the last line of the corresponding header file.

---

[1] The exact term would be "compilation unit."

Packages can use other packages, but you cannot declare a package within a package. In addition, while packages are still useful, they do have the same limitations as interfaces, as we will describe later in this chapter.

In SystemVerilog, enumerations and constants are in either compilation or $root scope. What that means in normal terms is that you should put enumerations, constants, and even parameters into a package.

> *Although packages have a number of unnecessary limitations, they are still a useful grouping construct.*

Separating HDL and testbench code

Have you noticed that the OOP features are different from the synthesizable subset of SystemVerilog? The sublanguages are different because the focus of HDL code is to create silicon, whereas the focus of verification code is to test that HDL code. HDL code is concerned with wires, nets, modules, and loads. Verification is concerned with class hierarchies, randomization, stimuli, BFMs, and checkers.

This handbook makes a distinction between the two code types, clarifying the reasoning behind when to use a given feature. You will not find modules in any of the code snippets or examples (other than in the HDL code and the testbench top). The authors feel that although the "language" may be the same, the goals are vastly different.

So how do you cross the border between HDL and testbench code? The only way SystemVerilog makes the connection between HDL code and testbench code is through an *interface*.

Wiggling wires: the `interface` concept

So what is an `interface` in SystemVerilog? At its core, an `interface` is a declaration of wires that logically go together. They are the "pins" of the chip, put into containers that make sense to your team. An `interface` is a SystemVerilog aggregation construct, similar to `module`, `program`, or `class`. An `interface` allows you to bundle a large variety of wires and registers into a single named entity that can be easily worked with.

For example, a protocol consisting of address, data lines, control, and clocking could form an `interface` as follows:

```
interface basic_inside;
  wire [7:0] address;
  wire [31:0] data;
  wire clock;
  wire address_latch_enable;
endinterface
```

So why use an `interface`? This is the *only* way to connect to "real" HDL wires. You could make an `interface` for every wire, but that would be clumsy and would not give a single name to related wires.

For each signal in an `interface`, you have a choice to make. The `interface` can either create the signal, or it can just refer to an existing signal in the testbench.

The `basic_inside` shown above defines all the signals as originating from inside the `interface`. Here is what an `interface` looks like when all the signals are passed in as parameters.

```
interface basic_outside ( //note the "(" instead of ";"
  wire [7:0] address, //use "," instead of ";"
  wire [31:0] data,
  wire clock,
  wire address_latch_enable
  ); //end of parameters
endinterface
```

When to use "inside" versus "outside" is both a matter of what language the HDL code is written in and your style. In our examples we have assumed that the HDL code is Verilog, not SystemVerilog, so it is more natural to have the testbench's `top.v` create the wires. Then the `interface` takes in all the wires, using the "outside" interface technique when it is constructed.

There are many variants of "inside" versus "outside," including having the `interface` generate the clocks (they can have initial blocks) and having the `interface` have some utility methods (they can have tasks and functions). An `interface` can be a good singleton (see the OOP Classes chapter) and can simplify the testbench top; it can also be a good way to stub modules. An `interface` can also be used to solve the age-old problem of when to sample and when to drive. On the downside, an `interface` cannot be "new'd," extended, or randomized, and it is cumbersome when used in an array.

Interfaces are THE way to connect the HDL code with the testbench code. Be cautious of using them for more than that.

Building and using interfaces

Okay, now that we can declare an `interface`, how do we create and use one? Interfaces can be created only in an HDL module or in a program block.[1] The authors prefer using a module, for several reasons. One is that we have a module of interfaces for each chip in the testbench. We can then either build a module with stub interfaces or a module with "real" interfaces, and the SystemVerilog testbench code is unaware of the change. The other reason we use a module is that we use the same program block for all of our testing. (This is explored in the Truss Basics chapter.)

Interfaces are made just like ordinary variables:

```
module real_interfaces;
  basic_outside outside_1 (top.adr, top.data, top.clk,
                           top.ale);
endmodule
```

To use an interface, you must put the keyword `virtual` before the interface name. The real interface will be passed in when the class is built:

```
class basic_bfm;
  function new (string name, virtual basic_outside bo);
    name_ = name;
    basic_outside_ = bo;
  endfunction
  extern task write (bit [7:0] ad, bit [31:0] data);
  local string name_;
  local basic_outside basic_outside_;
endclass
```

------

[1] Creating interfaces is a little weird, because of both the language's immaturity and the vendors' lack of conformance.

　• • • • • • •　　　　　

In the program block, you would create a `basic_bfm` and pass it a real instance of the interface:

```
program a_program;
  initial begin
    basic_bfm a_basic_bfm;
    a_basic_bfm = new ("First BFM",
                       real_interfaces.outside_1);
    a_basic_bfm.write ('h100, 'h02192007);
    //other code for the test...
  end
endprogram
```

The actual reading and writing of the HDL values is quite straightforward:

```
task basic_bfm::write (bit [7:0] ad, bit [31:0] data);
  @ (posedge (basic_outside_.clock);
  basic_outside_.address_latch_enable <= 1;
  basic_outside_.address <= ad;
  @ (posedge (basic_outside_.clock);
  basic_outside_.address_latch_enable <= 0;
  basic_outside_.data <= data;
endtask;
```

All these steps may seem a little confusing, but once you've done this a few times, it becomes formulaic. The far trickier parts of verification are the OOP parts.

First create a module to hold the "real" interfaces of the chip, then connect these interfaces to your BFMs, monitors, and so on, in the program block.

Summary

This chapter wove together three themes: the evolution of OOP, the evolution of verification, and the way SystemVerilog expresses OOP features.

We spent some time looking at the class declaration, with its accessor labels, constructors, data members, and tasks and functions.

We then took a look at inheritance and its two main techniques: inheritance for implementation, and inheritance for code interface.

We discussed packages as a way to avoid name collisions. We pointed out the usefulness of packages, as well as their warts.

Finally, we talked about interfaces, *the* way to connect testbench code with HDL code.

For Further Reading

- Again, we invite you to read *SystemVerilog for Verification: A Guide to Learning the Testbench Language Features*, by Chris Spear.

- The web has a some good introductory information about SystemVerilog. Some good sites are www.doulos.com and www.asic-world.com.

- The official standard for SystemVerilog is IEEE 1800. While dry reading, it is a must read, at least for the basic data-type sections. It is also a required reference for what the compiler might be complaining about.

A Layered Approach

C H A P T E R 4

*It is tempting, if the only tool you have is a
hammer, to treat everything as if it were a
nail.*

Abraham Maslow

For longer than we know, humans have organized themselves into
layers. From the family and tribe all the way up to national governments,
we have created roles and responsibilities. Closer to the hardware domain,
both VHDL and Verilog also use a layering concept, employing entities
or modules to break up a task. The software domain uses the related
concepts of procedures (*methods*) and data structures (*classes*). A reason
we humans make layers, with associated roles and responsibilities, is to
simplify our lives.

This chapter looks at how using layers can organize the task of verifying
a chip. We look at a generic chip, albeit one with a "System-on-a-Chip"
bias, and come up with a set of standard, well-defined layers, roles, and
responsibilities. We leave this chapter with definitions of standard verification layers and detailed diagrams of functional "boxes" and how
they are interconnected. Part II of this handbook will show a fully
implemented SystemVerilog environment that uses this approach. Part III

will talk about general object-oriented programming techniques for implementing these classes and connections. These techniques, applicable to most of the languages used for verification, express the reasoning behind the layered approach discussed below.

Overview

Throughout this chapter, little distinction is made among architecture, design, and coding. This is because these activities are interrelated, and occur at most stages in a project. Also, even with the initial architectural efforts, you should have a plausible implementation in mind; otherwise, the architecture may create problems when you are coding.

At many layers, verification environments tend to have the same set of problems. The essence of this chapter is to show how these common problems can lead to common solutions. By reusing solutions, the team can be more productive.

Specifically, this chapter covers the following topics:

- The importance of code layers, roles, and responsibilities

- How to go from a whiteboard verification system to classes and interconnects, using a standard framework

- Some common components, roles, and responsibilities of a verification system

There are many successful hardware products. Because success demands more success, the hardware produced in the next revision of a product will be more complex than the current version. In addition, the sales staff wants the product in the shortest possible time. The three competing factors of quality, functionality, and time to market create stress on the verification team. You are expected to produce more in less time—*and* with increased quality.

So how do you do that? You could add members to your team. While it is certainly true that there is an appropriate number of people for every task, adding people creates several issues. One is the need for increased communication; adding a team member increases the need for each member to interact with the rest of the team, decreasing productivity.

Another issue with adding members is team dynamics; each time new people are added to a team, it takes time for the team to become fully productive again. Finally, there is the fact that a well-integrated team can outperform an average team by a huge margin.

Okay, so adding people is a difficult way to build a quality verification system faster. The authors believe that a good way to do more quality work in less time is to increase productivity. As humans have done in the past, productivity can be increased by using layers. Now, we are not saying a government is a superefficient operation, but rather that a small team can be more efficient if the verification system is divided into layers. In addition, the resulting system is more likely to be simpler and able to be "warmed-over" for the next project.

> *A major tenet of this handbook is that the most productive individuals and teams use a layered approach.*

By using layers to separate the tasks of verification, common techniques and solutions can be seen. This allows the team to build up a library of standard solutions to common problems. Each of these solutions can be given a name (sometimes as a *base class*), along with a defined role and responsibility.

In this chapter we use layers to create a verification system. Starting with a whiteboard block diagram, we define layers, roles, and responsibilities and (in theory) arrive at a well-designed system. This technique is used to show how to move a verification system quickly from a whiteboard block diagram to classes and functional code.

We do not talk much about language specifics in this chapter, because the technique of using layers is applicable to almost any language. The next part of this handbook, Part II, shows specific implementations in SystemVerilog.

A Whiteboard Drawing

Most verification systems start on a whiteboard or something similar. Some engineers get together and discuss how they are going to test some part of the chip or maybe the entire new product. This initial effort results in an understandable and "clean" block diagram. However, transforming this whiteboard sketch to a similarly clear code architecture and implementation is difficult. This chapter outlines a layered approach to this transformation.

Note that the layering process occurs in one form or another at many levels of a verification system, from the full system level down to individual functional blocks. In addition, the classes and code are constantly refined and modified as the project progresses, so the use of these techniques is both fractal and recursive. This section focuses on this OOP process at the outermost level—in other words, from a system perspective.

A top-level whiteboard drawing might look something like this:

Whiteboard Drawing

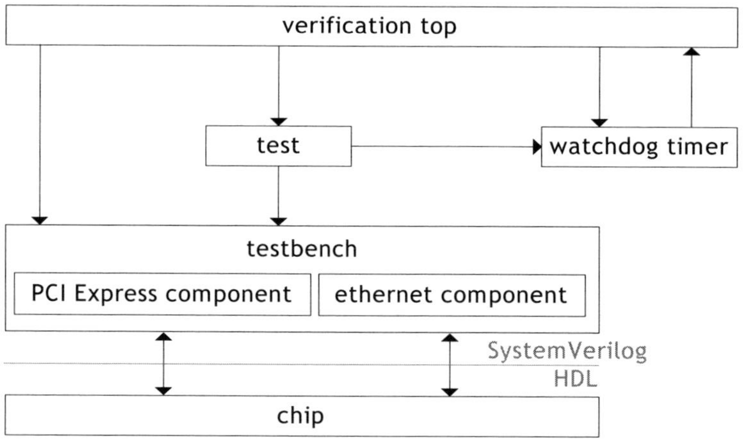

- The *verification top* block is responsible for instantiating and ordering the events of other components.
- The *test* block is responsible for controlling and synchronizing each component.

■ The *watchdog timer* is a time-out module that ends the simulation should something unexpected occur that would make a test run forever.

■ The *testbench* block is responsible for instantiating each *component* block. The PCI Express and Ethernet components here are important, because they transform test commands into bus transactions on the chip. They include methods for data generation and randomization, as well as drivers and monitors, as will be discussed in greater detail below.

An "ends-in" approach

So where should you start after the first whiteboard drawing is done? A common approach involves starting at the lowest level of abstraction (the connection layer) and coding the next layer up, continuing upward until every layer—including the test layer—has been designed. This is called the "bottom-up" approach.

However, there is another approach. This alternate approach still starts with the lowest layer of abstraction (because this is the best-defined layer), but then builds the top layer next and saves the middle layers for last. This approach is called an "ends-in" design, because you start by working on the connection layer and the top-most layer, and then build your way inward. This is the authors' preferred approach, because it maximizes what you already know. You know what the connections of the chip are (or at least most of them). You have an idea what a standard test looks like. You can therefore build these layers so that they are "reasonable" (that is, others—on the team or off—can reason intelligently about them). It's then an engineering effort to make trade-offs between complexity and adaptability, in order to connect the ends together into a system that is reasonable at all layers. This is not simple, and it requires a lot of experience, but the next part of this handbook contains a working example of how to do it. Part III of the handbook discusses techniques to evaluate the trade-offs.

Refining the whiteboard blocks

It would be tempting to define a class for each block in the whiteboard drawing shown in the preceding section. While it is possible to do so, this is not a good solution, because each block, especially a component block, contains too much functionality to fit well into a single class. Having classes that are too large leads to a brittle and often complex design that is not adaptable, or even maintainable.

Instead, it is a good idea to look closer at each block and define another set of layers. This makes sense, because most blocks can have several well-defined abstraction layers. This is what the rest of the chapter will address. Each major section below introduces the general roles and responsibilities of a block or abstraction level. The sections even get a bit more specific, suggesting common names for classes at each level. Some of these names are already common in our industry.

The "Common-Currency" Components

The first step in transforming a whiteboard diagram to code is to focus on the chip connections. The authors call the set of resulting classes the *connection* layer. In the whiteboard drawing above, there was a PCI Express component and an Ethernet component. Because these component blocks cover a lot of functionality, there needs to be a set of classes for each block, as discussed below.

The resulting classes are an example of a design pattern that the authors call the *common currency* of a verification system, because they are used so frequently. In fact, the classes are an implementation of the common-currency pattern, because the chip can have most of the components running for every test. Put simply, common currency can be considered a concept or pattern, of which component classes (such as PCI Express and Ethernet) are specific instances. These classes are the "money" of the verification system's "economy." Every team member should be able to identify the currency—in other words, the roles and responsibilities of the various connections of your chip.

There are many ways to identify a common-currency class in an OOP language. One way is to have the class inherit from a common-currency base, such as `class pci_express_monitor extends monitor`. In this case the `monitor` base class has a set of methods that `pci_express`[1] is expected to implement. Another way is to use a naming convention, such as `class ethernet_monitor`. Note the absence of the base class. While the "monitor-ness" of the class is not enforced by the compiler, you can bet the team will have expectations about what this class does.

Sometimes this naming-convention approach is best if the base class has no methods, or has just light-weight ones such as `start()`, `stop()`, and `report()`. The art of deciding what is a class, a convention, or a base class is up to you. Part III of the handbook discusses the various options and trade-offs.

The Component Layer in Detail

As mentioned above, each component can be divided into more layers and classes. This promotes adaptability and makes sense, because a component straddles abstraction layers; at the highest abstraction layer it consumes transactions, and at the lowest it wiggles wires.

The approach used by the authors is to break each component into three sublayers. The lower a class is in the component, the more the chip details that are handled. This layering process is a technique to manage complexity by allowing higher-level code (such as generators or monitors) to describe the problem in a more abstract way, thus providing a simpler code interface to the tests and making them both clearer and more portable.

The following figure shows how the component layer is in turn broken down.

---

[1] Or any other class that extends the monitor base class.

Component Classes

generator | checker
transaction

generator agent | checker agent

BFM agent | driver agent | monitor agent
agent

BFM | driver | monitor
connection

virtual interface(s) — SystemVerilog / HDL

chip

There are three abstraction layers.

- The *transaction* layer consists of fairly high-level classes, such as generators and checkers.

- The next layer down in detail is the *agent* layer. This is the layer that implements the connection policy and converts between high-level transactions and low-layer method calls.

- The lowest layer is the *connection* layer, in that the objects in this layer drive and sense the chip wires.

Let's look at the connection layer first.

The connection layer

The most detailed layer of the common-currency classes is where the monitor, drivers, and bus functional models (BFMs) exist.[1] This is shown in the highlighted section of the following figure.

[1.] With multilayered protocols, the fractal nature of a layer must be considered. Depending on the test to be run, there will be monitors, drivers, and so on at each level of the protocol.

Component Classes – Connection Layer

In this handbook, monitors and drivers are considered one-way connections,[1] while BFMs are considered two-way connections. These classes are generally the only ones that drive or sense the wires.

The connection-layer classes are complex and have a broad footprint. In other words, they have lots of methods, encompassing everything you want to exercise. The classes are extremely portable, because, by definition, the protocol from one chip to the next is well-specified. If, on another project, you have that same protocol, the monitor/driver/BFM from the connection layer should be easily adaptable.

The connection-layer classes have only a set of task and function methods. Their role is to take procedure calls and execute the wire dance that is specified by the protocol. These classes are responsible for the mapping between a method call and wire-change sequencing. Whether, and in what order, these methods are called is the concern of the next layer.

---

[1] A driver sends data and a monitor receives data. Note that the driver or monitor may both drive and sense wires to do this function.

The agent layer

The next layer up is called the *agent* layer, as shown here:

Component Classes – Agent Layer

The agent layer is responsible for using various connection-layer classes to implement the upper layer's requests. It is called the agent layer because it acts as a go-between for two relatively well-defined components. Commonly, the classes in this layer add some sort of queue, for data or control actions, depending on what the upper layers generate or check.

This layer may also have several implementations. For example, many chips have multiple ways to send the same data. There could be register, FIFO, and DMA ways to interact with a chip. You could have three different connection classes, one for each of these methods. The test could randomly pick which method to use, and would still look the same.

Because the agent layer is the transaction layer's view down into the chip, it also implements the connection policy. For example, you could use a simple direct connection, thus forcing the generator and driver to act in tandem. Alternatively, you could implement a multipoint connection, using events or other broadcast mechanisms, to connect several drivers to a single generator. The same concepts can be used for the monitor-to-checker connections.

The transaction layer

The uppermost layer is called the *transaction* layer, as shown in the following figure.

Component Classes – Transaction Layer

The transaction layer uses the previously discussed layers to exercise some component or feature of the chip and validate the response. The exercising (or driving) part of this layer is called a *generator*. The response validating (or receiving) part is called a *checker*. Note that there may be more than one generator or checker if different types of traffic are to be exercised on a component. There is a trade-off between making a single, flexible and capable generator or checker, and having several, fixed-function simple classes. Choosing which to use is a judgment call for your team.

These three layers are portable code and can be used for almost any chip. Of course, the generators will have to be constrained, randomized, and started. Also, the checker will have to be waited on until it has checked all the expected chip responses. These activities are the responsibility of the higher test components, as will be described in later sections.

One interesting property of the common currency of component classes is that each generator and checker probably has at least one thread of execution. This is because hardware is massively parallel, and can operate multiple protocols independently. In addition, the rate at which the generation and checking occurs is only indirectly tied to the behavior of the chip's wires. For example, a single generated "packet" may require many bytes to be transferred at the wires, or several data bytes may be gathered from the wires before the checker is called.

So this is how a test component is broken down into classes that are manageable and adaptable. The process of examining each chip protocol, and then implementing a set of interacting common-currency classes to handle the generating/checking and driving/monitoring, is now repeated for each protocol. If this method of using layers and the underlying protocol is well-defined, then there is a good chance that these classes will be used again in later projects.

The Top-Layer Components

The whiteboard drawing is now pretty much converted to code for the chip protocols. Because we are using an "ends-in" approach, we will tackle the components at the top before we look at the middle layers.

The *top* layer has standard form, roles, and responsibilities, just as the component layer did. The following figure shows the top layer with its standard classes.

At the very top is the *verification top*, shown in the following figure. This component builds the other top-level components and sequences the initialization, randomization, execution, and shutdown of the simulation. It would be easy to mistake this for the test itself—but wait.

It is better to abstract the functionality of the verification top into an independent function. In this way, the specifics of the current project's tests and testbench are removed from the more generic build, startup, and shutdown sequence. The verification top can then be used on multiple projects.

Top-Layer Verification Components

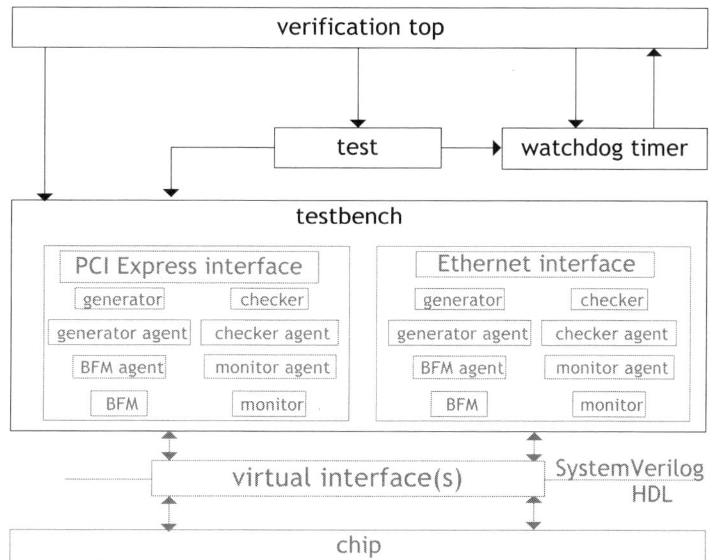

Also at this layer, the three main workhorses of the verification system are the testbench, the test, and the watchdog timer. Like the verification top, the watchdog timer is most likely a generic implementation. Its role is to shut down the system if too much time has elapsed.

The testbench is probably specific to a project, but it is the same for most tests. Its role is to contain the component objects and perform chip-wide initialization and possibly configuration. The test is responsible for constraining and sequencing the component objects. (This process is described further in the next section.)

Note that the test changes with each scenario you want to run. The testbench and test implementations differ between projects and runs, yet their code interfaces (by definition, the public class methods) remain constant. At this high level of abstraction, the concepts of building, configuring, running, and shutting down a verification test are uniform.

It's up to the team to decide how to design the class methods for these standard top-level classes, as well as how to design the build/configure/

run sequence for a simulation. The Truss Basics chapter has base classes for these standard classes.

What is a Test?

The previous section went quickly over the roles and responsibilities of a test. Because a test is an important concept in verification, let's be a bit more thorough. A test is one of the main top-layer classes. It is responsible for exercising some subset of the features of a chip while background traffic or other activity is occurring. The test's main partner is the *testbench*. Before we talk about the test, let's review the role of the testbench and see how the test uses it.

The testbench contains the connection-layer objects for each of the chip's protocols. In general, the testbench only *holds* these objects; it's up to the test to *use* them. However, there is a small exception: when the chip has mutually exclusive features or protocols. In this case, the testbench might have an object or two that "chooses" an appropriate configuration.

The test is responsible for "deciding" which features of the chip are to be tested. Because most chips are massively parallel devices, a well-designed test focuses on some part of the chip—but it also exercises other functions or protocols of the chip simultaneously. Often there is a primary protocol or feature to be tested, and a number of independent, secondary protocols or features.

After a test has "decided" on a protocol or feature to focus on, it must constrain the random behavior of those features. The test selects and writes a configuration to the chip, probably by interacting with the connection BFMs. After that, the test starts up all the component generators and runs them until some end condition is met and the end of the test is signaled. This end condition could be either elapsed simulation time or whenever the primary component exercise has completed. Finally, the test waits for all the other components and then reports success or failure.

Because a chip may have several protocols, it can become tedious and clumsy to work with the component-layer generators, BFMs, and checkers directly. The test may become cluttered with management code, and it may be difficult (for all but the original writer) to figure out the point of the test. Also, many tests will use many combinations of chip protocols and features, so that much of the code is replicated. For these reasons, it's better to group a chip's component-layer test into a concept the authors call a *test component*. In addition, the other protocols that are just exercising the chip as background traffic are packaged into *irritator* components. These components are "middle layers"—that is, they connect the connection layer to the top layer.

Here is an example of what the components of a simple PCI Express test might look like:

Example PCI Express Test

Main test part PCI Express test component

Background traffic Ethernet irritator

Background traffic USB irritator

Background traffic UART irritator

The test component is described in the next section, and the test irritator is described after that.

To summarize, a testbench holds the component-layer objects, which the test selects, constrains, and controls. It is good practice to break a test into test components, one for each protocol or feature of the chip. For a specific test, a few test components are the main components, while other components—the irritators—provide background traffic.

The Test Component

The whiteboard drawing probably does not include information about the middle layers. This section details some of the questions and decisions related to the middle layer. It is at this middle layer where common questions such as "What object should set what parameters?," "What should be randomized and when?," and "How do we know when we are done?" are answered. In some sense this is the hard part of the verification system, where a lot of mental energy is spent.

The implementation of the middle layer starts with listing the types of exercises you want to perform on a protocol or feature of the chip. Often these test requirements take the form of sequences that exercise the basic data paths and functionality of the chip, including error cases.

Once you have this list, you create a middle-layer class to represent each exercise. The authors call these classes the *test components* of a test. A test component does not just represent a stimulus or a scenario for a protocol; it also includes the end condition. In a sense, a test component has a code interface like that of the verification top, evidence of the fractal nature of a layered approach.

A test component is used to exercise some specific functionality of the chip. In fact, a test component is often used with other test components to create a rich test, with the other test components acting as "background noise" generators (irritators). This is a benefit of designing test components as a separate class, instead of directly implementing the exercise in the test.

Another benefit of separating the test from test components is that the test is a layer above the test components, creating them, giving them the appropriate parts of the testbench, and setting their parameters. This allows different tests to drive the same test components differently, perhaps letting a test component "roam" on its parameters, or maybe constraining it to hit a corner case.

In general, a test component, on construction, gets references to a generator, a driver/BFM, and a checker of a chip protocol. The references come from the testbench.

Why not take in the entire testbench? By not just referencing an entire testbench, but instead taking the pointers it needs, each test component manages complexity by minimizing the assumptions on the environment. In addition, the test component maximizes the chance that it can be adapted to other testbenches. By having a test component itself perform an exercise, instead of directly implementing the exercise in a test, you have a better chance of ensuring the adaptability of the exercise.

Let's look at a concrete example. Suppose you are driving packets into a chip from one protocol and collecting processed packets on another protocol. To test this data path, your test system will look something like this:

Test Component Connections

The packet test component class gets a pointer to both BFMs, a generator, and a checker. The role of the packet test component is to exercise some part of the chip by using these other components.

For some methods the test component may just relay calls from the test. In this example, a `start()` method calls `start()` on the generator, checker, driver, and monitor. Recall that the test could have called all the component layer objects directly; the test component layer just makes the test clearer.

One nonrelay task that the test component performs is to sequence the generator. For example, the packet test component mentioned above would probably tell the generator to generate a certain number of packets.

This number might be set by a randomized parameter, or it could be fixed. There may be other component-layer parameters that the test component controls, such as packet size or protocol configuration parameters. This is where the test component implements what is in the test plan.[1]

The test component is essentially an aggregator. Given pointers to a component's generator, BFM/driver, and checker, the aggregator sequences these.

The Test Irritator

We have only one more part to consider before we complete the conversion from a whiteboard drawing to a verification system. This last part addresses how to write background traffic components. Recall that a robust test has a test component and several other background traffic components. The idea is to ensure that the chip can function in a real-world scenario.

When you start writing tests, you will probably start with test components. Then, after the tests are stable, you'll want to add auxiliary test components.The name the authors use for these auxiliary components is *irritators*. An irritator is most likely to have "evolved" from one of your test components for that chip interface.

When converting a test component to a background traffic generator, you must alter the component so that it addresses not an internally governed amount of traffic but rather an externally controlled one—so that, for example, it does not use a fixed-length group of packets but instead an infinitely repeating sequence of packets. In other words, you want the nonessential irritators to continue doing whatever chip exercise they do until your test says to stop.

The Truss chapter has an `irritator` base class that is inherited from the `test_component` base class. If you write your test irritators so that

---

[1.] A discussion of a test plan is beyond the scope of this handbook, but basically the plan is a list of the exercises you need to perform on the chip.

they use these base classes, irritators can be implemented with very little effort.

By adding irritators, you can write tests that are understandable yet reasonably complex.

A Complete Test

Let's take a step back and look at what we have accomplished. We have progressed from refining a whiteboard drawing to defining responsibilities for classes and code. We now have definitions for sets of classes for each protocol of the chip. We have a testbench that contains instances of each of these classes. We also have a set of test components and irritators that can be combined like building blocks to create a diverse set of tests.

At the top-most layer are the tests. These tests should not only exercise a main function, but also leverage the work of the other team members by using irritators as nonessential traffic. These tests should exercise the chip fairly well.

Shown below is an example of a test that uses the layers we have talked about, an example UART test with several irritators added.

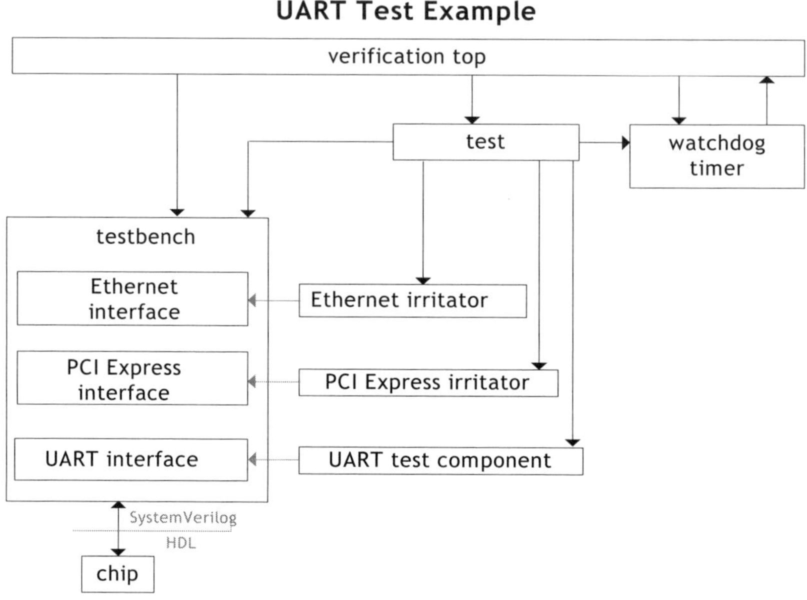

This example test includes a "main" test component, called *UART test component*, which probably walks a range of configurations and sends some amount of data. The test also includes Ethernet and PCI Express irritators.

Many choices still remain, such as exactly how to sequence the bring-up, running, and shutdown of these test components. The next part of this handbook provides a standard framework to help you make these choices.

You must still decide how to randomize and constrain the test parameters. For example, there are implementation choices regarding what control variables to include in the test components and the irritators, as opposed to control variables in the generator and configuration for the connection layer. Although these are not simple choices, the next few chapters should help to clarify the trade-offs.

This completes our first pass at converting from a whiteboard drawing to code. As mentioned earlier, this exercise is essentially repeated continually, even as code is written. In other words, reality happens when you write the code.

Summary

This chapter talked about layers. We talked about using layers to increase productivity, by managing the complexity of a verification system.

We talked about "ends-in" coding, where you start at the bottom and top of the test and code towards the middle. We considered this technique of looking at the chip and creating component layers as the first step in creating a verification system. We then went to the top layer, and talked about the verification top and the three top components: the watchdog timer, the test, and the testbench.

Next, we entered the middle layer, where we talked about using a test component to exercise a particular configuration or data path of a chip protocol. The idea that a test really should have several components exercised at once formed the reasoning behind the irritator layer.

We ended the chapter with a quick tour of a completed test, noting that there are still more decisions to be made as the implementation of the verification system proceeds.

For Further Reading

- *The Mythical Man-Month,* by Fred Brooks, is the classic handbook that talks about why one should not put additional people on a team to solve a problem. He argues, as we did in this chapter, for a more productive team.

- *A Few Good Men from UNIVAC, by* David E. Lundstrom, talks about the concept of a focused and productive team. This book is about the origin of supercomputers.

- *On the Criteria To Be Used in Decomposing Systems into Modules,* by D.L. Parnas, is a 1972 landmark paper on how to go from a problem statement to a design. The fancy name for this process is called *decomposition*. The current fancy terms for thinking about design are "design patterns" or "factory objects." However, be careful with these recent concepts; they refer to good high-level solution templates, but those templates must be applied with care and experience.

- The concept of *design directions* came from Harlan Mills and Niklaus Wirth at IBM in the 1970s. Their original idea was to use a "top-down" approach, but all variants have been popular at various times. The authors believe that an "ends-in" approach is the best for the class of problems encountered in verification.

Part II:
An Open-Source Environment with SystemVerilog

The previous part of the handbook was a high-level look at System-Verilog and how to architect a verification system by using layers. Now we focus on a specific implementation of such a system.

This part of the handbook introduces two open-source libraries, called Teal and Truss, that together implement a verification environment. The authors and others have used these libraries at several companies to verify real projects.

The libraries are free and open source because the authors feel strongly that this is the only way to unite and move the industry forward. Locking up people's "infrastructure" is not the way to encourage innovation and standardization—both of which are needed if the verification industry is to improve.

Consequently, you'll find no simulator-company bias in these libraries. These libraries work on all major simulators.

In this part we discuss the following:

- Teal, a set of utility classes and functions for verification

- Truss, a layered verification framework that defines roles and responsibilities

- How to use Teal and Truss to build a verification system

- A first example, showing how all the parts we talk about fit together

Teal Basics

Coming together is a beginning. Keeping together is progress. Working together is success.

Henry Ford

Building a verification system is a daunting task, but build we must. That is why we use the technique of layering, to break the problem down. By starting with the lowest layer—that is, the one that directly drives and senses the wires—we can start to get some real work done. That is covered by SystemVerilog's interface feature. The next layer is the basic building blocks, such as loggers, parameters, and memory access. Teal is a collection of building blocks. Teal tries to be as unobtrusive as possible, allowing you and your team to make framework decisions as you see fit.

This chapter introduces Teal and shows how to use it. We'll talk a bit about the main parts of Teal—for example, how you can get and set memory or registers in the chip, and how you can create a flexible, yet simple log messages.

Overview

Teal is a set of utilities implemented in a package. Teal is tiny, consisting of only a handful of source files, yet it provides the necessary minimum features for verification. (Teal is freely available at www.trusster.com.)

Teal is unobtrusive; it does not get in the way of your verification structure. Teal provides the general functions that most verification systems use.

The authors realize that many companies have developed their own version of a "Teal." We encourage those companies to contact us and share their experiences, so Teal can be made better. This is one of the reasons why Teal is open source.

Teal's functionality provides the basis for functional verification, but it serves as only a small part of a verification project. You must still write code that stimulates the design, checks the output, and controls the randomness. That is the real work of a verification project. (The next chapter talks about an open-source verification environment.)

Teal's Main Components

It is important to decide on a "common currency" when designing a class library. The rest of this chapter describes the common currency of the Teal system—that is, the fundamental building blocks of Teal-based verification.

The following is a summary of the most important classes and functions of Teal; more detail is given in the following sections.

- *The* vout *class*—This Teal class is used for logging, to help trace what happens during a simulation. The vout class provides the ability to report, for example, debug, error, and other informative messages in a consistent format that is coordinated with HDL outputs.

- *The* vlog *class*—This class is a global resource that coordinates all the logging from your code. It receives all vout messages from the simulation and implements a filter chain, so you can add useful features such as replicating output to a file and removing messages or parts of messages.

- *The* memory *functions*—These functions provide an abstract interface for reading and writing memory. Internally, a group of memory banks are used to handle memory read and write requests, providing great flexibility.

- *The* vrandom *class*—Because using random numbers for test values is a staple of modern verification, this class is Teal's stable random-number generator. Though SystemVerilog provides a sophisticated rand/constraint capability, this class is small and completely under your control. The class provides independent streams of stable random numbers that are guided by a single master seed. Of course, the numbers all have their own seed as well, based on what you provide. This allows these numbers to be stable and create a new stream of numbers only when you change their local seed or the master seed.

- *The* dictionary *functions* —These functions are a global service that abstracts how to set parameters in your test. They provide the functionality to get and retrieve parameters from code, the command line, or external "scenario" files.

- *The* latch *class*—This class provides a latchable/resettable event, so that you do not have to worry about thread execution order, or about missing an event that occurred during previous forks. You'll use this whenever you have two classes that want to communicate a status, such as a generator or monitor, to a higher level to indicate completion. The examples show several uses of a latch.

All of these classes and functions are described in the following sections.

Using Teal

It's time to dive into some details regarding how Teal can be used for functional verification. This walk-through of Teal makes it easier to understand the "real world" examples presented in subsequent chapters, while illustrating how Teal can be used in your environment.

A simple test

When the simulation begins, your program block is executed. The program block could be as simple as the following:

```
'include "teal.svh"
program verification_top();
  initial begin
    teal::vout log = new ("first code");
    log.info ("Hello Verification World");
  end
endprogram
```

Logging Output

Because a lot of debugging is done by reading simulation log files, in order to see a progression it is important to organize simulations well. In other words, to enable postprocessing, error counting, messaging, and possibly filtering, it is important to have a consistent message format. Fortunately, the logging facility in the Teal classes encourages such uniformity. Teal comes with a standardized, customizable logging mechanism, called `vout`.

Teal uses a two-level logging scheme, as shown in the following figure. In any code that needs to print information, a `vout` object is created. As many `vout` objects as needed can be created—which is good, because each `vout` object can have a relevant instance name.

Each `vout` object "under the covers" calls a global service `vlog` object. This is done so that there is a single point of control where the reordering, changing, or deletion of parts of any message can be done.[1]

vout and vlog Objects in Teal

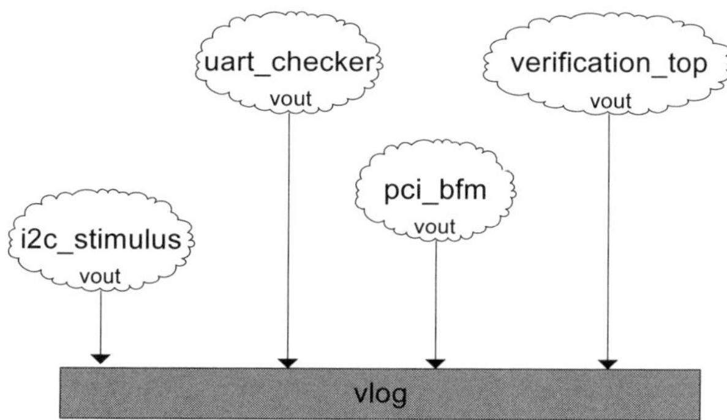

Note that `vout` simply takes a string. By calling the appropriate function—`info()`, `error()`, `fatal()`, `debug()`—`vout` can provide this information to the `vlog` class. This information is not just sent as a string, but is passed as metadata that can be useful for further processing. This is described more fully below.

When you create a `vout`, you give it a string that represents the functional area it is in. You can then build any number of message statements. For example, note the following:

```
'include "teal.svh"
program verification_top();
  initial begin
    teal::vout log = new ("a test");
    log.info ($psprintf2 ("val 0x%0x", 207218);
  end
endprogram
```

This example prints the following (assuming a simulation time of 77 ns):

```
[77 ns] [a test] val 64 0x32972
```

---

[1] Although describing this capability completely is beyond the scope of this handbook, subsequent chapters show several examples.

[2] Note that while `$psprintf()` is not a part of the standard, all the major vendors provide this call.

Note that when you call one of the display methods—`info()`, `error()`, `fatal()`, `debug()`—the vout instance adds the simulation time and the functional area to the message, then sends the message to the `vlog` global service. It does not send the message as a text string, which would not allow the efficient modification of the message; rather, it sends the message as a set of pairs of IDs and strings. This allows you to instruct the `vlog` instance to modify messages with respect to their components—for example, to demote errors to a warning, or stop all output from a file or a functional area.

However, you often do not need to use the global filtering mechanisms of `vlog`. Instead, you can turn off the display of parts of a message directly, at the `vout` instance. This is done by calling the `message_display()` function with the ID of the item you want to display (or not). By default, all items are displayed.

Most verification systems have several levels, or types, of messages. Teal, being no exception, uses the following general categories:

- `info(msg)`— Used for standard messages.

- `debug(msg)`/`debug_n(<level>, msg)`—Used when a test wants to display a little more diagnostic information. This is a level-sensitive output; the `vout` class has level-setting methods and accepts a level for debug messages. The message is displayed only if the level of the message is less than or equal to the level that is set. The `debug(msg)` method uses a level of 1.

- `error(msg)`—The error type is used when the chip's actual behavior is different from the expected.

- `fatal(msg)`—This more-severe error type ends the simulation after displaying the message.

Examples of the above are provided in later examples.

Using Test Parameters

It is often important in functional verification to provide test parameters. These are frequently used, among other things, as bounds for random tests. For example, a single test case may have several different sets of test parameters—including, say, the maximum number or errors before the test is forced to terminate, or the maximum amount of time the test is allowed to run. Sets of test parameters can also be used to direct a test into interesting corner cases.

Because such parameters are commonly used, Teal provides a standard, flexible way of working with them. Test parameters can be defined by means of text files, code, or command line entries. Teal handles simple integer and string parameters as well as complex parameters.

Teal's `dictionary` functions are used to access test parameters. Teal maintains a list of parameter names and values, so that a test, for example, can query the dictionary and recover the value.

When you call the `dictionary_read(string)` function, Teal reads a text file, takes the first word on each line as the parameter name, and saves the rest of the line as data for that parameter. A special keyword, `#include`, is used to open other files from within files. If a parameter is repeated, the last definition is saved.

In addition to using files, you can also use code to add parameters. When you do this, you have the option of replacing an entry or not.

Parameters can also be entered on the command line. In this case, they override any parameter set by a file or the code. In this way, a parameter can have a default value but still be overridden by a script.

As an example, let's suppose we are testing a UART interface. We have a default parameter file that sets up default constraints, and then each specific test overrides a few values as well as defines its own parameters. The default parameter file could look like this:

```
//in default_parameters.txt:
force_parity_error 0
dma_enable 1
baud_rate 115200 921600
```

A specific parity-error test case could use the default parameter file and override the `force_parity_error` setting like this:

```
//in parity_error_test_parameters.txt:
stop_error_probability_range 32.81962 75.330
#include default_test_parameters.txt
force_parity_error 1
```

The `#include default_test_parameters.txt` line above tells the dictionary to open the `default_test_parameters.txt` file. The `force_parity_error 1` repeats the `force_parity_error` parameter and overrides the default value.

It is not always appropriate to use files to pass parameters. Using files can be good if you need to have many different test parameters and a few basic tests. However, it can be clumsy to make sure the files stay with the respective test code. Therefore, the examples later in this handbook use the code mechanism. Nevertheless, including such files, or even passing parameters on the command line, can be done after most of the test is written, without having to modify the test itself.

So how do we pick up the parameters? The following is a complete basic example of how these parameters could be retrieved:

```
'include "teal.svh"
import teal::*;
program verification_top();
  initial begin
  //reads file shown above
    dictionary_read ("default_parameters.txt");
    vout log = new ("first_parameter_example");
    log.info ($psprtintf ("force_parity_error is %0d",
          dictionary_find_integer ("force_parity_error")));
  end
endprogram
```

Because most parameters are not strings, Teal provides a function, `find_integer()`, to convert parameters to the correct variable format. The `find` function always returns a string—either an empty string (`""`) if the parameter is not found, or the actual string associated with the parameters. The `find_integer()` returns an integer if the string is found, or returns the passed-in default if it isn't.

If using `find()` or `find_integer()` is not appropriate for your application, you can use the `$sscanf()`. This allows the code to create a stream from a string, from which you can then extract the `chars`, `ints`, `doubles`, and so on, as needed.

For example, to read the `stop_error_probability_range` (from the example above), you would use the following:

```
'include "teal.svh"
import teal::*;
program verification_top();
  initial begin
    dictionary_read ("parity_error_test_parameters.txt");
    //reads "32.81962 75.330" from file into s
    string s = dictionary_find("stop_error");
    real stop_error_min = 0;
    real stop_error_max = 0;
    int foo = $sscanf (s, "%f, %f", stop_error_min,
                       stop_error_max;
    teal::vout log = new ("showing double double reads");
    log.info ($psprinf ("Stop error range is %0f to %0f",<<
                        stop_error_min, stop_error_max));
  end
endprogram
```

Accessing Memory

For most verification projects it is important to be able to access memory. Sometimes you want to do this in zero simulation time. Allowing "backdoor" accesses of memory improves simulation performance, allows the monitoring of memory for automatic checking, and makes it possible to insert errors into memory for test purposes. Teal provides such a "backdoor" mechanism but also, of course, supports "front-door" access, which can map some memory address ranges to a transactor-based model.

Teal defines each accessible memory (transactor model or memory array) as a `memory_bank` object. A `memory_bank` object can be accessed directly through member functions called `to_memory()` and `from_memory()`, but each memory can also be associated with an address range, through the `add_map()` function. In this way, memory can be accessed through addressing by means of `read()` and `write()` functions.

Working with address ranges has many advantages, because it creates code that is easier to understand.

When writing a memory transactor, you must define your own `memory_bank` object, using virtual interfaces or pointers or channels to BFM transactors.

The following example shows how HDL memory arrays can be associated with an address range and accessed. An example of how to write memory transactors is in the UART example chapter.

A memory example

The following diagram shows a small part of a larger testbench structure. This environment verifies a graphics chip that saves graphical texture information in its memory cache. In order to speed up simulation, back-door loading of the texture into the chips memory is used.

Memory Bank Objects

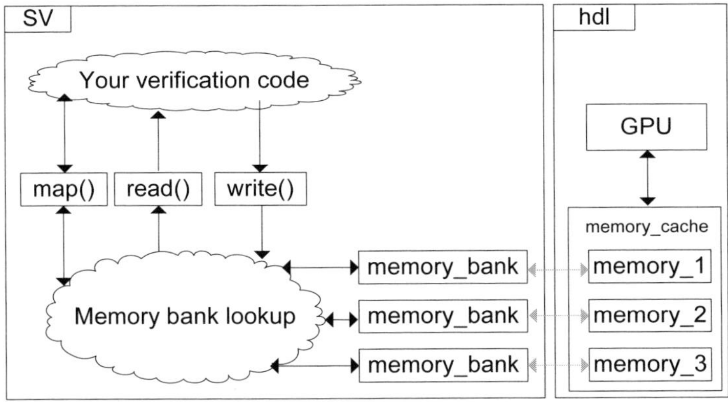

To support direct memory access you need to do a few things. First, you need to create an interface to the memory register bank. Then you need to define a subclass of `teal::memory_bank` that takes in a virtual interface of that type and performs the `from_memory()` and `to_memory()` functions. After that, you need to create an instance of that class and give it to Teal's memory manager by means of the `teal::add_memory_bank()` function. Then you need to define an address range for each memory instance to be used, allowing Teal to translate from an address to a specific memory. Finally, once the address range is established, you access that memory through `read()` and `write()` functions. Whew! That sounds like a lot of work, but it really isn't. It's just that describing code is clumsy.

To do this in the environment pictured above, an interface and a memory class is defined (that is, the memory model is instantiated as `memory1`, `memory2`, and `memory3`, above). So part of the memory model looks like this:

```
//In Verilog HDL, here is what the actual memory arrays are
reg[31:0] memory_bank_1[1024:0]; //Actual memory array
reg[31:0] memory_bank_2[1024:0];
reg[31:0] memory_bank_3[1024:0];
//now in a SystemVerilog header file,
//perhaps chip_interfaces.svh, define the interface and
// class
interface gpu_ram (output reg[31:0] bank[1024:0]);
endinterface
class gpu_memory_bank extends teal::memory_bank;
  function new (virtual gpu_ram g);
    super.new ("gpu ram");
    gpu_ram_ = g;
  endfunction
//now override the access tasks
  virtual task from_memory (bit [63:0] address,
                output bit [teal::MAX_DATA - 1:0] value,
                input int size);
    value = gpu_ram_.bank[address];
    log_.info ($psprintf ("Read[%0d] is %0d", address,
                            value));
  endtask
  virtual task to_memory (bit [63:0] address,
              input bit [teal::MAX_DATA - 1:0]  value,
              input int size);
    gpu_ram_.bank[address] = value;
    log_.info ($psprintf ("Write[%0d] is %0d",
                            address, value));
  endtask
```

```
    local virtual gpu_ram gpu_ram_;
endclass
```

Now the real interfaces of the chip must be created. The way the authors like to do this is in a separate module from the `top.v` (testbench top). The reason for this is to support building "dummy interfaces" if parts of the chip are not built yet or are stubbed out for some reason.

```
module interfaces_dut;
   gpu_ram gpu_ram_1 (top.memory_cache.memory_1);
   gpu_ram gpu_ram_2 (top.memory_cache.memory_2);
   gpu_ram gpu_ram_3 (top.memory_cache.memory_3);
endmodule
```

As can be seen in the code below, the memory model gets instantiated three times as `memory1`, `memory2`, and `memory3`. In the `verification_top()` program, the three memories get address ranges declared like this:

```
program verification_top();
  initial begin
    gpu_memory_bank gpu_memory_1 =
            new ("top.dut.gpu.memory_cache.memory_1",
                interfaces_dut.gpu_ram_1);
    gpu_memory_bank gpu_memory_2 =
            new ("top.dut.gpu.memory_cache.memory_1",
                interfaces_dut.gpu_ram_2);
    gpu_memory_bank gpu_memory_3 =
            new ("gpu_memory_3",
                interfaces_dut.gpu_ram_3);
    memory::add_memory_bank (gpu_memory_1);
    memory::add_memory_bank (gpu_memory_2);
    memory::add_memory_bank (gpu_memory_3);
    memory::add_map
        ("top.dut.gpu.memory_cache.memory_1",
                'h100, 'h200);
    //The following assumes the subpath memory_2
    //is unique
      memory::add_map ("memory_2", 'h201, 'h400);
      memory::add_map ("memory_3", 'h401, 'h600);
  end
endprogram
```

Now any test can access these memory spaces through simple read and write function calls. Furthermore, neither reading nor writing memory consumes any simulation time.

So why bother with all this machinery? First, most of the machinery would have to have been built anyway. SystemVerilog needs a `virtual` interface if it is actually going to poke the HDL. So you need to define a virtual interface and its real instance. Then you need to have the assignment statements that are within the memory bank. The benefit that the memory bank and map machinery add is to abstract the pin manipulation, or reach inside the DUT, with a simple read/write interface. Once memory banks are in place, this interface can be used to access any part of the DUT. The test writer is freed from figuring out how to perform access, and simply needs to look at the register/memory map specification for the chip in order to read/write. The power of this simplicity cannot be overstated.

Note that this technique can also be used to simplify register access. The examples use this technique, so you can see how it makes the driver code easier to write and understand.

In addition, should the testing team want to switch to front-door access, such as going through PCIe or a I2C protocol, this can be done without any changes to the "user" code.

A basic memory access would look like this:

```
teal::uint32 actual;
teal::write('h10a, 22, 32); //'ha in memory_1 = 22
teal::read('h10a, actual);
if (actual != 22)
  begin
    teal::vlog log = new ("memory_example_1");
    log.error (
      $psprintf
      ("At memory_1['ha] got 'h%0x expected 'h%0x",
        actual , 22));
  end
```

Note that while the memory is written and read at `'h10A`, the actual memory is accessed at `'h0A`. This is because of the `add_map()` that we performed, which allows the rest of the verification system to read and write memory as specified in the chip's memory map. Teal takes care of finding the correct memory bank to access, and then removing the mapping offset.

Constrained Random Numbers

It is important to have a stable, repeatable, seeded random-number generator. While SystemVerilog provides an extensive random feature and a constraint language, sometimes more is not always better. For example, not knowing what SystemVerilog uses as the local seed can lead to a test that changes its random behavior when you do not want it to. The constraint language is another consideration. While vast, it is still declarative, and it's up to the constraint solver to pick among the universe of acceptable random numbers. This can lead to confusing behavior and contradictions in the solver. Finally, sometimes it's easier to express the constraints on the random behavior as just a procedure. The examples in the book sometimes use Teal's random and sometimes SystemVerilog's, just to show that a problem can be solved several ways. As usual, it's up to your team to decide what it wants to use.

Teal's `vrandom` class provides independent streams of random numbers that can be initialized from a string or file. There are also convenient macros for the most common random calls, such as getting a random integer value or getting a value from within a range.

The rest of this section describes the required initialization of the random generator and some simple examples.

Required initialization

Before using any random numbers you must initialize the random-number generator. This is done by calling the `init_with_seed()` function and passing it a 64-bit seed value. It is recommended that higher-level code keep track of this seed value and pass it to the random number generator. To initialize the random seed generator you would call the following:

```
teal::uint64 master_seed;

...

// master seed gets initialized by higher layer
```

```
teal::vrandom_init_with_seed(master_seed);
```

After the random-number generator is initialized, it is ready to be used. The examples in this handbook use the dictionary to get the master seed. Also, the examples use the same seed as SystemVerilog, so there is only one master seed.

Using random numbers

Because integers are so commonly used here, Teal provides a couple of macros to deal with integers. After you have initialized the random-number generator you can call these macros directly. The most often used macros are RAND_32 and RANDOM_RANGE, which generate a 32-bit random value and a 32-bit value within a range, respectively.

Here are some examples:

```
'include "teal.svh"
program verification_top();
  initial begin
  teal::uint32 a_rand32; 'RAND_32(a_rand32)
  teal::uint32 a_random_range;
        'RAND_RANGE(a_random_range, 0, 'h030837);
  teal::vout log = new (" random number test");
  log.info ($psprintf (
          "a_rand32 is %0d a_random_range is %0d",
          a_rand32, a_random_range));
  end
endprogram
```

When you want to create more-elaborate random numbers, you need to work with the vrandom class directly. The vrandom class is a simple class that you can draw numbers from after it is created. This gives you more direct control over the generation of random numbers. The base vrandom class provides a uniform distribution, but you can create your own classes to have segmented, logarithmic, or other distributions.

You would create an object for your inherited class and draw a number like this:

```
my_vrandom a_random = new ("some string", some_integer);
teal::uint32 a_random_value = a_random.draw();
```

These parameters are hashed with the master seed and are used to initialize this particular random-number generator. You may want to pass in your own values.

Working with Simulation Events

SystemVerilog provides events to allow threads of execution to communicate. Unfortunately, there are some restrictions on the event handling. One of the most tricky is that the events do not latch, so if an event happens and a receiving thread is not waiting, the event never "happened.[1]" This can be desirable, but most often is not what one wants. So Teal provides a latchable event.

The `latch` class just includes a bit with the event, so that the "tree falling" event is remembered. When another thread goes to wait on the event, it will either pause the execution and wait for the event, or, if it already occurred, just return.

The latch can be "one-shot," that is, automatically resetting after being read, or it can wait for a manual clear. Sometimes it doesn't matter, if the event will happen only once in the simulation. This is generally the case with checkers and their "done" event.

```
'include "teal.svh"
program verification_top();
  initial begin
      teal::latch latch_1 = new ("latch for checker");
      fork begin
        begin
        $display ("do checking");
        #10;
        latch_1.trigger ();
        end
        begin
        #11; //thread "misses" the event
        latch_1.pause ()
        $display ("checking is done");
```

[1.] Like the tree falling, unheard, in the forest.

```
        end
      end
    join
    $display ("test is done");
  end
endprogram
```

Now obviously this example is contrived, but the point is that sometimes you cannot guarantee that a thread is waiting before the event is issued. It's safer to use this little object than to debug a race condition that might change with the seed.

Later in this handbook there are many examples of how to use latches.

Summary

This chapter introduced an open-source package called Teal. We talked a bit about what Teal provides and its components.

Logging is a very important capability of a verification system. Teal's `vout` class and the global service class `vlog` provide a uniform, yet very flexible, logging capability.

Almost all tests need to have control parameters set by code or files. Teal's dictionary provides a global service for managing parameters.

The `memory` functions of Teal can be used for both register access and internal chip memory accesses. If reads and writes are extracted from the actual underlying mechanism, different transactors can be used.

Random numbers are essential in verification systems. Teal provides the `vrandom` class, a stable, independent random-number generator.

We ended the chapter with a look at Teal's `latch` class, and considered why you might need to use it.

Truss: A Standard Verification Framework

> *Truss, and verify.*
>
> *Anon.*

Have you ever watched a building being constructed? Early in the project, when the frame of the building is just a skeleton, it's not clear what the finished building will look like. However, as construction continues, from the windows down to the cubicles that are our workplaces, the intent of the framework becomes clear. In fact, a large part of the building's presence depends on the fundamental structure.

This same basic process occurs when we build a verification system. Early in the project, the application framework is built. The result of years of best practices from both the verification and software fields, Truss is an application framework for verification. It is an *implementation*, and therefore makes some decisions about how things should be structured. With verification as with construction, the framework sets the tone for the system.

Truss is a layered architecture, so you can choose how to implement the layers. Although it makes very minimal assumptions, Truss does provide some base classes and conventions as a guide.

Overview

This chapter presents three main topics:

- The roles and responsibilities of the various major Truss components

- How these components work together

- How you adapt this framework for your verification system

This chapter builds on the two previous chapters of the handbook. It implements an open-source verification infrastructure based on the discussion in the Layered Approach chapter. It also uses the Teal library described in the previous chapter.

Teal provides the fundamental elements of a verification system and supports a wide array of methodologies. Truss, on the other hand, provides the infrastructure layers above Teal, adding a set of classes, pseudo-templates, idioms, and conventions to facilitate the construction of an adaptable verification system.

One of the tricks in building a reasonable infrastructure is to find the *key algorithm*. The rest of the algorithms can usually fit around that key algorithm. For example, in a video editing program the key algorithm is all about getting the pace of the edits right. When you watch a movie, that happy, sad, or scared feeling you get comes from how well-timed and precise the changes in scene are.[1] The authors, having developed software for video editing systems, know that in this domain the key algorithm is implemented by adjusting the edit points of a few seconds of video while the video is constantly looping around those edits. This is not a trivial thing to do, because multiple streams of video and audio, possibly with software algorithms to implement effects, are changing as the user is adjusting the edit points.

[1.] Okay, emotions also come from the music, but everything works together.

So what's our point? Well, in the verification domain, the key algorithm is the sequencing of the various components of the system. The authors refer to this as "the dance," as there are usually a few interacting components involved. As we talked about in the Layered Approach chapter, the top-level dance takes place between the test, the testbench, and the watchdog timer. Truss implements this dance in the `verification_top` program—but Truss does not stop there. The authors believe that this dance is the key algorithm in several layers of the system, so we created a `verification_component` base class. Also, we created `test_component` and `irritator` base classes to be the "top" at the component layers of the system. Recognizing and reusing the dance is a significant part of Truss.

This chapter explains the major components of Truss, providing code examples where appropriate. Subsequent chapters provide more-detailed examples.

General Considerations

The authors have worked on several different implementations of verification systems before Truss was available. While at a high level verification systems can be described uniformly, the language used to build them has a lot to do with how a specific framework is constructed.

SystemVerilog considerations

SystemVerilog is an evolving standard, and vendor compliance with the current standard is also evolving. However, Truss can be used with multiple vendors. Although it's possible to build a "better" OOP-based verification framework by supporting only one vendor, single-vendor support would have made Truss a verification novelty. Truss represents an appropriate cross-vendor SystemVerilog framework that many years of software and verification experience can create.

SystemVerilog does not have copy constructors, or operator overloading. However, it does provide object-instance reference counting, and Truss relies on this by passing around objects liberally. SystemVerilog does not have a concept of interface versus implementation. It does allow `'include` and `extern` on methods, which Truss uses to create the illusion of interface files. Truss also uses pure virtual functions to communicate a code interface.

Not all vendors support static methods, so Truss uses a function in a package to implement a singleton. SystemVerilog does not allow task calls in functions, so Truss has a fair number of `void` functions.

The point of the above is to show that Truss is carefully architected to minimize the impact of vendor and language constraints, while providing a basis for verification efforts.

Keeping it simple

A stated goal with both Teal and Truss is to avoid unnecessarily complicated code. SystemVerilog has many legacy and new features, but many times they are not appropriate. It is easy to get distracted with language techniques and forget that the real goal is to keep the whole team productive.

For example, implementing a generic interface for a verification component, such as a transactor, as a template can be tricky. Sometimes using a template can be more complicated than simply replicating code.

Sometimes only a convention should be used. An example of this is the *generator* concept. One could define a virtual base class, yet the common methods come down to just `start()`, `stop()`, `report()`, and a few others. It turns out that this concept of `start()`, `stop()`, and so on is common to a large set of verification tasks, and is represented in Truss as the virtual base class `verification_component`. However, the concrete subclasses are inherited from `verification_component` only if they use the bulk of the methods. Any smaller subset uses the same named methods as a convention instead.

> *In this way, the framework is not warped to fit a generic class. Even more important, your design is not warped to fit the generic class.*

Truss implements a specific methodology for functional verification. As in any endeavor to generalize, the terrain is fraught with peril. Nevertheless, as writing code entails making judgments about what is the "right" decision, Truss attempts to generalize a style of verification. Deciding on the right balance between generic and specific is a judgment call for the team. The idea behind Truss is to foster a small, usable, portable, and adaptable methodology for beginners through experts. As such, Truss provides an example of the techniques presented in Part III of this handbook.

Major Classes and Their Roles

Truss is an implementation of the layers talked about in the Layered Approach chapter. Consequently, there are only a few top-level components—the verification top, the testbench, the test, and the watchdog timer. Each component has a specific role. These components and their roles have been architected to allow a large amount of flexibility with a relatively simple interface. These top-level components (and those the next level down) are shown below:

Verification Component Hierachy

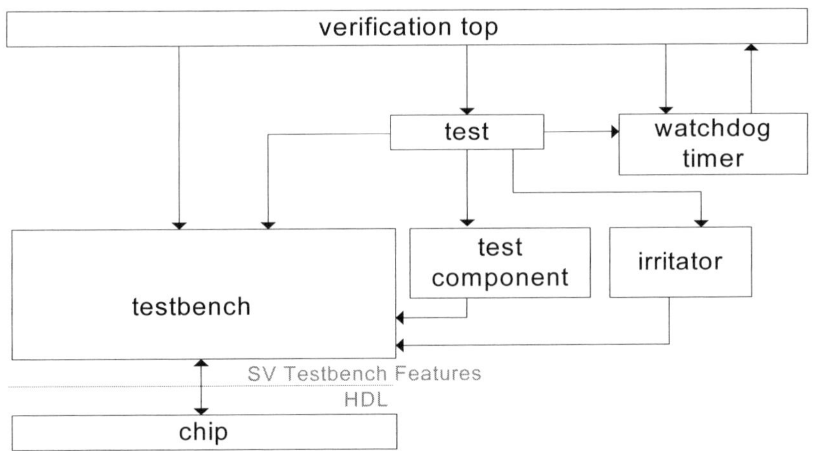

The top-most component is the `verification_top()` program, whose role is to create and sequence the other components through a standard test algorithm. (The algorithm is explained in detail in the next section.) In addition, `verification_top()` initializes all global services, such as logging, randomization, and the dictionary.

The watchdog timer is a component created by `verification_top()`. This component's role is to shut down a simulation after a certain amount of time has elapsed, to make sure the simulation does not run forever.

The testbench top-level component is the bridge between the System-Verilog verification world and the HDL chip world. As such, the testbench's role is to isolate the tests (and test writers!) from having to know how transactors, traffic generators, monitors, and so on interact with the chip. Whether a bus functional model (BFM) writes to registers or forces wires should not be of concern to the test writer.

In addition, the testbench holds the configuration objects of the chip. This is needed by the BFMs, transactors, and similar agents to be able to configure the chip correctly. There is probably a configuration object for each protocol of the chip. For chips that contain internal functions, such as direct memory access (DMA), there may be a configuration object for each function.

The last, but certainly not the least, top-level component is the test itself, whose role is to execute a specific functionality of the chip. It does this by using the testbench-created BFMs, monitors, and generators. The test is responsible for choosing among the testbench's many configurations and capabilities and exercising some subset of the chip's functionality. In general, the test contains very little code. This is because any code it contains may need to be used in other tests as well. To support code that is more adaptable, a test normally consists of several test components, as will be discussed later. The exception is for directed tests, in which case registers may be overwritten, specific traffic patters sent, or specific corner cases exercised directly in the test component.

Key test algorithm: The "dance"

The top-level components of the previous section have a complex, yet necessary, set of interactions. This ensures the maximum flexibility for a test, while providing a known set of interactions. This is one of the tricky parts of a verification system. This section discusses this standard algorithm, which we call the "dance."

In general, the top-level components are created, randomized, and then started. Then `verification_top()` waits for the test and testbench to be completed. This is called the "polite" path. If the watchdog timer decides that a timeout has occurred, the "impolite" path is taken and the simulation ends.

The order of these calls can be better visualized on an event diagram, as shown below. The four columns show the main components. Execution starts at the top left line, and the arrows represent function calls to the other components.

The Dance

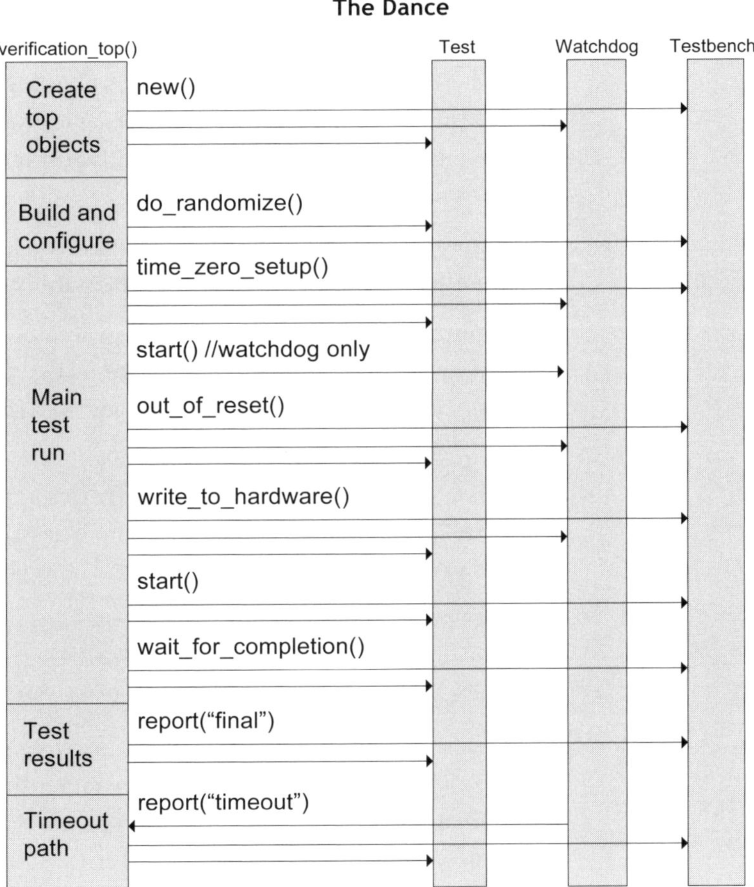

The first thing that `verification_top()` does is build the global logging objects. These provide for logging, as well as for shutting down the simulation after a threshold number of errors have been logged. (See `truss_vout.svh` in the index.)

Then `verification_top()` reads the dictionary file (if it exists). This is to allow the test constraints file to override any default settings put there during the construction of the test, the testbench, and their subordinate components.

Then, after the global logging objects have been created, and the dictionary read, `verification_top()` allocates the top-level objects. The test is given a pointer to the testbench, so that it can interact with the testbench. It is also given a pointer to the watchdog timer, in case a part of the test wants to force a shutdown or override the `verification_top()` default timeouts. The watchdog is given a pointer to an object so that it can call the final report method with a "watchdog timeout" string prefix.

At this point, all the top-level objects are constructed. As part of their construction they are expected to have established default constraints.

After initializing the random-number generator, `verification_top()` calls `test.do_randomize()`. Once the test is randomized, then `testbench.do_randomize()` is called.

At this point, it is expected that the test and testbench have built their respective subcomponents and are ready to run the test. The first step is the `time_zero_setup()` method, which is used to force wires and initialize interfaces prior to bringing the chip out of reset.

As expected, the next step is `out_of_reset()`, which is used to bring the chip out of its reset state and set it for initialization through the backdoor or register writes.

The next step, `write_to_hardware()`, is where the BFMs are called to initialize the chip. This can be done by either the test, the testbench, or a combination of the two. What is appropriate depends on your situation, as discussed in subsequent sections.

At this point the system is ready for traffic flow. The `start()` method directs the testbench and test to start running. The testbench is started first, to allow monitors and BFMs to start, followed by the watchdog timer. Finally, the test is told to `start()`, which generates the actual traffic.

Next, `verification_top` calls `wait_for_completion()` on the *testbench*. If your design makes the testbench aware of what checkers are in use, this call waits for the testbench checkers to complete. If not, this method simply returns.

Then `verification_top` calls the *test*'s `wait_for_completion()`. If your design makes the test aware of what checkers are in use, this call waits for them to complete. (This is the style used in the examples.)

At this point, the test is almost finished. The testbench and test are called to report their final status.

Then `verification_top()` checks to see if any errors were reported. If none were reported, the test is considered to have passed. It may seem weak to accept that the absence of errors is sufficient to consider a test passing. In practice, however, there is no other choice. At the top level, one must trust that the lower-level objects do their jobs. Note that this usually means that in-flight data must be weeded out as the checker proceeds.

Now if the watchdog timer triggers, a different path is taken. The watchdog immediately calls the report method on `verification_top`. Note that the watchdog itself uses an HDL-based timeout, so that if the report method hangs, the simulation still ends.

The verification_component Virtual Base Class

While the test and the testbench are completely different classes as far as their roles and responsibilities are concerned, their code interface to `verification_top` is the same. For this reason a common class was created. This common class, used as a base for both the test and testbench, is called the `verification_component`.

The `verification_component` is a virtual base class. As such, it provides pure virtual methods for the dance described in the previous section. In addition, `verification_component` provides a constant name and a logger. The interface for `verification_component` is shown below.

```
package truss;
    typedef enum {cold, warm} reset;
  virtual class verification_component;
      protected teal::vout log_;
      protected string name_;
      extern function new (string n);
      'PURE virtual function void do_randomize();
      'PURE virtual task time_zero_setup();
      'PURE virtual task out_of_reset(reset r);
      'PURE virtual task write_to_hardware();
      'PURE virtual task start();
      'PURE virtual task wait_for_completion();
      'PURE virtual function void report(string prefix);
      function string name();
  endclass
endpackage
```

Although `verification_component` is a base for the test and the test-bench, it is also useful as a base for other objects.

Detailed Responsibilities of the Major Components

The previous sections discussed the roles of the major components and how they were sequenced to run a test scenario. This section dives down a level, discussing in more detail the specifications of the major components. (Because `verification_top` was discussed in detail in the previous section, it is not discussed further here.)

The testbench class

The `testbench` class has two main responsibilities. One is to isolate the test writers from the actual wire interfaces. The other is to provide "one-stop shopping" for all the generators, checkers, monitors, configuration objects, and BFMs/drivers in the system. The reason to put all of your components into a single object is to facilitate the adaptation of components into multiple tests. In this way, a test writer can see all of the possible "building blocks" that are available.

The `testbench` class can be a passive collection point for all these components, or it can play an active role in bringing the chip out of reset, generating traffic, and knowing when the test is done. In theory, only the global functionality should be handled by the testbench. For example, the testbench probably should bring the entire chip out of reset, while the test can bring separate functionality out of reset. In practice, the test and the testbench share the work.

In general, it is better to let the test or test components control the simulation. This is because a test or test component can then be adapted for several different types of tests.

A more active testbench may, as a counterpoint, simplify a large number of tests in a way that a test base class cannot, because the testbench has direct access to all the chip's wires.

Understand that the more test knowledge a testbench has, the more all tests must act the same or have control over that testbench's functions. This can be good or bad. The specific responsibilities for control and functionality—test or testbench—are, of course, up to the verification team.

As an implementation detail, Truss provides only a `testbench_base` class. What `verification_top` builds, however, is a `testbench` object. You must provide a `testbench.svh`. This file declares a `testbench` class, which should be inherited from `truss::testbench_base`. You will probably also have a `testbench.sv`, which contains the implementation.

The testbench is passed a helper object, called `interfaces_dut`. This is because, in SystemVerilog, classes are not allowed to make cross-hierarchical references. What does that mean? It means that the testbench can access the chip wires only through an interface. The `interfaces_dut` class is where all those interfaces reside!

But each chip will have different interfaces, so how can we make a generic program that builds a specific testbench? The answer is to have you write a function, called `build_interfaces()`, that is called by `verification_top`. Truss creates the virtual base class.

```
virtual class interfaces_dut; endclass
```

Then you define a function to build your specific derived class.

```
function interfaces_dut build_interfaces ();
```

Note that the truss `testbench_base` constructor receives this pointer as the base class.

```
virtual class testbench_base
   extends verification_component;
   function new (string n, interfaces_dut dut);
endclass
```

In your testbench you must downcast[1] the `interfaces_dut` to recover the derived class you created in your `build_interfaces()`.

```
interface alu_input (
  output reg   [31:0] operand_a,
  output reg   [31:0] operand_b
  );
endinterface
class interfaces_alu extends truss::interfaces_dut;
  virtual alu_input alu_input_1;
endclass
  //MUST be in file build_interfaces.svh!
  function interfaces_dut build_interfaces ();
    interfaces_alu alu = new ();
    alu.alu_input_1 = real_interfaces.alu_input_1;
  endfunction
  class testbench extends truss::testbench_base;
    function new (string n,
              truss::interfaces_dut dut);
```

---

[1] Generally, downcasting is a bad thing, but we cannot think of a better solution given the context.

```
        interfaces_alu alu_dut;
        super.new (top_path, dut_base);
        //Note: Downcast to recover pointer
        truss_assert ($cast (alu_dut, dut_base));
    endfunction
  endclass
```

The authors generally declare the virtual interfaces of a chip in a file called `interfaces_<chip_name>.svh`, and then put the instances of the "real" interfaces in a file called `interfaces_<chip_name>.sv`.

Finally, these "real" interfaces must be in a module called `real_interfaces`. This is because some vendors require you to name all the top-level modules in a simulation, and the truss script uses that module name.

The authors realize this is a fair amount of structure. The good part is that you would probably need most of this structure anyway. We are just providing a naming framework for these methods and classes so that a generic program and script can be used.

> **You must implement a function called** `build_interfaces()` **in a file called** `build_interfaces.svh`. **You must also have a top-level module called** `real_interfaces`.

Watchdog timer

The watchdog timer component is responsible for providing an "impolite" shutdown if the test has executed for too long. The timer has two timeout mechanisms: one triggers when the watchdog HDL timer triggers, and the other triggers after the first trigger has occurred.[1]

The watchdog timer uses the dictionary to get its timeout values, which are sent to the HDL on `time_zero_setup()`. The `start()` method starts

---

[1] The watchdog timer is simple in theory, but often hard to execute correctly. To be sure, it must have a clock and a countdown time, but even this basic level can be problematic. Should you use wall clock time, simulation time, or both? Should the HDL timer be internal or external? What resolution should it have? Should the test be able to extend or communicate the expected time of the run?

the timers. The HDL watchdog uses an internal timer. If it were to use a passed-in clock, that clock may inadvertently be shut off.

Once either timer triggers, the watchdog HDL timer is notified and a second timer is started. If this timer expires, `$finish` is called. This might happen, for example, if there is some code in the report that is still reading registers, but the chip is unable to respond.[1]

After the watchdog is notified of an HDL timeout, the `report()` method in `verification_top` is called. This allows the test to report which checkers have completed and which have not, helping to provide a clue as to why the simulation ran too long.

> *The watchdog interface must be put in your derivation of the* `interfaces_dut` *class. In addition, a real interface must be created and passed to the watchdog in its constructor.*

Test class

The test class is responsible for selecting, configuring, and running all the appropriate generators, BFMs, monitors, and checkers. It is also responsible for selecting the configuration of the chip to be used.

While you could directly implement the above responsibilities in the test class, Truss encourages another style. In Truss the test is intended to consist of a number of independent, smaller components called *test components*. These components are the ones that actually do the work; the test's role is to create, constrain, configure, and sequence the components, as appropriate for the test at hand. The reasoning behind having multiple independent components is that this is close to the real operation of the chip, where each feature is expected to operate simultaneously. In reality, the chip has common resources that must sequence or arbitrate the use of features. It is in these common resources where the more tricky bugs lurk.

---

[1] The authors worked on a project where the final report code read the status registers to make sure that functional area of the chip did not have any errors. However, when we added a power-down test irritator, the read hung the system. It took us a while to find the offending code.

Using this method, the test's direct responsibility is to map the features of the chip (as presented by the testbench's data members) to a set of classes inherited from the `test_component` base class. The test would then add constraints to adapt the test component to the test at hand, as in the following example:

```
class ethernet_basic_packet extends truss::test_base;
  local ethernet_test_component ethernet_data_1_;
  local ethernet_test_component ethernet_data_2_;
  local pci_irritator pci_express_1_;

  function new ethernet_basic_packet(testbench t,
                                     truss::watchdog w);
    ethernet_data_1_ = new (t.e_generator_1, t.e_bfm_1,
                       t.e_checker_1);
    ethernet_data_2_ = new (t.e_generator_2, t.e_bfm_2,
                       t.e_checker_2);
    pci_express_1_ = new (t.pci_generator_1, t.pci_bfm_1,
                     t.pci_checker_1);
  endfunction
  task time_zero_setup();
    ethernet_data_1_.time_zero_setup();
    ethernet_data_2_.time_zero_setup();
    pci_express_1_.time_zero_setup();
  endtask
  task out_of_reset(reset r);
    ethernet_data_1_.out_of_reset(r);
    ethernet_data_2_.out_of_reset(r);
    pci_express_1_.out_of_reset(r);
  endtask
  task write_to_hardware ();
    ethernet_data_1_.write_to_hardware();
    ethernet_data_2_.write_to_hardware();
    pci_express_1_.write_to_hardware();
  endtask
  task start ();
    ethernet_data_1_.start();
    ethernet_data_2_.start();
    pci_express_1_.start();
  endtask
  task wait_for_completion ();
    ethernet_data_1_.wait_for_completion();
    ethernet_data_2_.wait_for_completion();
```

```
            pci_express_1_.wait_for_completion();
        endtask
          function void report (string prefix);
            ethernet_data_1.report(prefix);
            ethernet_data_2.report(prefix);
            pci_express_1.report(prefix);
          endfunction
          function void do_randomize ();
            ethernet_data_1.do_randomize ();
            ethernet_data_2.do_randomize ();
            pci_express_1.do_randomize ();
          endfunction
        endclass
```

In the above example, the `ethernet_basic_packet` test uses three test components, two of which are identical. It connects up the appropriate testbench objects and forwards to every test component the following test calls:

```
time_zero_setup(), out_of_reset(), start(),
wait_for_completion(), do_randomize(), and report()
```

So why do testing in this more complicated manner? In addition to the previously mentioned idea of simulating close to real-world conditions, an important reason is to maximize the adaptability of the test components. In the example above, we used the same test component for both Ethernet ports. Also, when the test components take in only the parts of the testbench that they need, they (1) make explicit what they are using, and (2) minimize the assumptions on the rest of the chip. This, as will be highlighted in the single UART example in Part IV, allows a test component to be reused for other chips that have only a subset of the original chip's functionality.

Test components are critical to the adaptability of a verification system. In general, the test components themselves do not know whether they are running in parallel with other test components or are part of a series. Thus, the most adaptable components are these test components, as will be discussed further in the following sections.

As an implementation trick, `verification_top` builds a test by using a `define` called TEST. This trickery, set up by the `truss` run script, allows the script to compile in a different test, while leaving the rest of the build

   • • • • • • •

image the same for all tests. This allows each test to be its own class (inherited from `test_base`). This cleverness helps one avoid a bad experience in the future. Assume that your team had written on the order of 50 tests, and then a new test was created that required a new subphase to be added to the dance. Although the other tests did not need this new method, you cannot add the default method. This is because all the tests are implemented as a test class. There is only one header `test.svh`, and 50 different `test.sv` files. By defining a base class, and then having the actual test be an inherited class (with a different header file), one can add methods to the base without affecting the existing tests.

There is one more part to a test that needs to be discussed. Often a test is made better by the addition of random background traffic. This traffic, be it register reads and writes, memory accesses, or just the use of other interfaces, can uncover corner cases, such as bus contention, that would not be found otherwise.

These background-traffic test components are called *irritators* and inherit from the `test_component` class. They differ from the standard test component in that they continue their traffic generation until told to stop by the test. Test components, by contrast, decide themselves when they are done, as determined by specified metrics, such as a stop time or the number of packets to send. (Irritators will be describe in more detail later in this chapter.)

With background traffic irritators, the test is written essentially as before. The exception is that the `wait_for_completion()` of the test calls the primary test components' `wait_for_completion()`. When the primary component returns, the test calls `stop_generation()` on all the irritators and waits for them by means of their `wait_for_completion()`. Then the test returns control to `verification_top`. (This is explained further in subsequent sections and in the examples in the chapters that follow.)

Test Component and Irritator Classes

As discussed in the previous section, test component-based design is central to a Truss-based test system. The authors have found that separating the test scenarios into test components has maximized the adaptability of the system. By using test components and irritators, test writers have been able to minimize their assumptions and distractions and concentrate on exercising the chip. Furthermore, other test writers can adapt what was done in other functional areas and inherit irritators (if they are not already present) for use as background traffic.

This section describes the responsibilities and interfaces of the `test_component` and `irritator` virtual base classes.

The test component virtual base class

The `test_component` is an virtual base class whose role is to exercise some interface of the chip. As discussed above, this functionality has traditionally been included in the test. The `test_component` describes the interface that all concrete implementations must follow.

In fact, you may have several types of `test_component` for a single interface, for example, a register read/write one, a basic data path one, and an error case one. The fact that these different exercises implement the same interface simplifies reasoning about them.

In practice, most test components use a generator and a connection-level object. Sometimes they may also be given a checker, depending on the designer's intent.

The `test_component` class is a `verification_component`, and has all the same phases. The `test_component` breaks down some of the `verification_component` methods into finer detail, as one would expect of a lower-level object.

Below is the interface for the `test_component` base class.

```
package truss;
  virtual class test_component
                    extends verification_component;
        function new (string n);
```



```
        task void start();
        task void wait_for_completion();
        function void report(string prefix);
    //protected interface
      'PURE protected virtual task start_();
      'PURE protected virtual task
        run_component_traffic_();
      'PURE protected virtual task start_components_();
      'PURE protected virtual task do_generate();
      'PURE protected virtual task wait_for_completion_();
      protected bit completed_;
  endclass
endpackage
```

As before, the methods `time_zero_setup()`, `out_of_reset()`, and `write_to_hardware()` are provided to allow the test component to interact with a BFM or driver. Note that a different, but equally valid, architecture would keep the connection-layer components private in the testbench and sequence them by means of the top-level dance. This assumes that the testbench knows what subset of the BFMs, drivers, and monitors, to start up.

The `start()` method is used to start the `test_component`'s generator, BFM, and so on. This method is implemented by a Truss utility class called `thread`. A `thread` class runs another virtual method, `start_()`, in a separate thread or execution. This allows a test class to do the obvious thing and just call `start()` on all the test components the test uses.

Let's look at the `start_()` method, as it is the main starting point for an interface of the chip. The `start_()` method runs two methods: a `start_components()` pure virtual method, and a virtual `run_component_traffic_()` with a default implementation. The idea behind the `start_components_()` method is that you call `start()` on your generators, BFMs, and so on, as appropriate. (The examples part of this handbook contains examples of `test_component`.)

The default `run_component_traffic_()` method calls `do_randomize()` (to randomize the test component and its components), and then calls `do_generate()`. In your `do_randomize()` method, randomize the data members that will be used by `do_generate()` to cause some traffic to be generated. In your `do_generate()`, take these data members and make the appropriate calls to the generators in the testbench.

An AHB example

An example might make the roles a little clearer. (Remember that there are several fully implemented examples in Part IV.) Suppose you are creating a test component to test an AHB[1] arbiter. The test component acts as a master, generating read and write requests to a number of slaves.

The generator in the testbench can generate a burst of reads or writes to a given slave, using a specific burst length. Assume that the generator has a channel interface that can take in an AHB transaction object. The randomize function of your `ahb_test_component` might look like this:

```
task ahb_test_component::do_randomize();
  burst_length_ = generate_burst_length(min,max);
  is_read_ = generate_type(min_type, max_type);
  slave_ = generate_slave(min_slave, max_slave);
endtask
```

The corresponding `do_generate()` might look like this:

```
task ahb_test_component::do_generate();
  //addresses are picked by the generator
  generator_.queue_burst (
    new (burst_length_, is_read_,slave_));
  done_.signal();  //Signals that test_component is done
endtask
```

Notice that by nature these calls are executed in a one-shot manner. That is, together they perform a single transaction. This is useful to allow an `irritator` to inherit from this test component later, to sequence this pattern any number of times and possibly change the randomization constraints as well.

So why have two separate methods?

By separating the randomization from the generation phases, one can inherit different classes that either (1) have different randomization characteristics (for example, logarithmic distributions of the burst length, or a pattern); or (2) send the data through a filter first, then to the generator.

[1.] AMBA (Advanced Microcontroller Bus Architecture) high-performance bus.

So now that the transaction has been generated, what should the `wait_for_completion()` method do? Because the generation is occurring in another thread, there should be a condition variable to communicate when it is done.

So the code might look like this:

```
task ahb_test_component::wait_for_completion_();
  done_.pause();
endtask
```

Test-component housekeeping functionality

The `test_component` class also provides a basic housekeeping bit that tracks when you return from the `wait_for_completion_()` method. This allows the `report()` method to determine whether you have considered the work of the component to have been completed or not. This can be very useful in a timeout situation, to see which components have not completed.

What you decide to do in the `wait_for_completion_()` depends on how you view your `test_component`. One view is that it is a traffic generator only, which can complete when the generation of traffic has been queued. It is then up to the testbench or test to determine when the chip has processed all the data. This will most likely involve a checker or monitor.

Another view is that your `test_component` represents a generate and check path through the chip. In this case, the completion of `test_component` signifies the completion of the entire exercise. (The examples in this handbook use this view.)

> *As always, the team must decide which view is better for their project.*

The irritator virtual base class

As discussed above, the `test_component` is set up as a one-shot traffic generator. This works for tests that are directed, and for tests where the completion event is predetermined—that is, tests that know before the `start()` call what the end conditions are.

However, sometimes it is not good design to have the `test_component` determine when completion is achieved. This is the case when, for example, you want to achieve a certain metric, and the measurement is not appropriate information for the `test_component`.

For example, you may want to send 100 bursts of some AHB traffic. While this could be included in the `ahb_test_component`, you might not want to measure completion by 100 bursts all the time. Instead, you might want to write a test that looks at the number of hits each slave device gets, and stop the test when all slave devices have been targeted. As another alternative, you might want a test to run until some goal is met; a goal could be any of the previous goals, or could involve some internal state in the arbiter.

The `irritator`, inherited from `test_component`, is used for situations such as these. The interface is shown below.

```
package truss;
  class irritator extends test_component;
    extern function new (string n);
    task stop_generation(); generate_ = 0; endtask
    extern virtual protected task start_();
    extern virtual protected task
      run_component_traffic_();
    extern virtual protected bit continue_generation();
    'PURE virtual protected task inter_generate_gap();
    local bit generate_;
  endclass
endpackage
```

The irritator overrides the `run_traffic_()` method of the `test_component` base class. It sets up a loop, calling the one-shot randomization and generation in the `test_component`'s `run_component_traffic_()` methods. The implementation is shown below.

```
task void irritator::run_component_traffic_();
  while (continue_generation()) begin
    super::run_component_traffic_();
    inter_generate_gap();
  end
endtask
```

The method `continue_generation()` just looks at a bit, which is toggled to 0 by a call to the `stop_generation()` method. This allows an external class to stop the continual loop of randomization and generation.

Note that there is a new virtual method in the `irritator` class, called `inter_generate_gap()`. Because the irritator is continually generating traffic, you might need a delay mechanism to prevent the generator from flooding the chip.

There are many ways to get this delay. For example, in one solution the generator and attached BFM/driver could execute the generate request as soon as it is called and thus take simulation time. In another solution, the way to get a delay would be to have a fixed-depth generator and BFM/driver channel.[1] This would put back-pressure on this generate loop. In still another solution, the generator could have a delay in clock cycles before returning.

Any of the above solutions is acceptable, but there is yet another choice. That option is to have the irritator itself provide the delay mechanism. The `inter_generate_gap()` is a virtual method allowing you to implement an irritator-based delay. This allows the irritator to decide on the throttle mechanism. Different subclasses could implement different policies. For example, an irritator could wait for a variable number of clock cycles. Another example would be to measure some parameter on the checker (such as packets in flight).

As always, the team must decide what is appropriate.

---

[1] This method is supported in Truss's `channel` class.

Using the irritator

The irritator continues this generate/wait loop until a `stop_generation()` is called. But how do you decide when to stop the irritator? The answer, of course, is "When the test reaches its goal." One goal could be that the "main reason" for the test has been achieved. For example, you can have the main goal be a test component, perhaps one that generates a fixed, but randomized, number of packets through a particular chip interface. The global goal in this case would be for the test component to achieve completion. Here is how the test code might look:

```
task noisy_packet_test::wait_for_completion();
  basic_packet_exerciser_.wait_for_completion();
  //'for_each is a macro from the quad_uart example
  'for_each(irritators_, stop_generation());
  'for_each(irritators_, wait_for_competion());
endtask
```

Ignoring the nontrivial constraining, selecting, and creating of the test component and irritators, what is accomplished in a few lines of code is a shutdown sequence that is powerful, while being a fairly simple idiom.

Note that a verification team could decide to use only irritators in their implementation. In that way, when to stop the test can then be determined by looking either at a checker or possibly at elapsed simulation time. The complex part of the test would then become the randomization and selection of irritators. The authors have worked on a variant of this methodology, and the resulting verified chip was a first silicon success.

Summary

This chapter introduced Truss, an open-source application framework.

We revisited the benefits of an OOP language such as SystemVerilog. We also stressed the need to keep things simple despite the features of this language, to avoid writing code that is unnecessarily complicated.

We talked about the key algorithm of verification, which the authors called the "dance." We showed how the dance is used by the `verification_top` program to run a test. We discussed the roles and responsibilities of the test, testbench, and watchdog timer, the main parts of the top-level dance.

We discussed the `verification_component` virtual base class, which provides pure virtual methods for the dance.

We then discussed the `test_component` and `irritator` classes, including their responsibilities and interfaces.

Truss Flow

C H A P T E R   7

Have you ever bought and assembled a piece of furniture from IKEA? In the store most of their furniture looks very simple, but when you get it home and try to assemble it, you realize that it's built from several smaller and often confusing pieces. Even with IKEA's famous assembly instructions, showing the "intent" for each piece graphically, assembly can still be confusing. Imagine how hard it would be without instructions.

The authors have had to learn many verification environments through the years, and this has often been a very confusing experience. What seems like a great concept with a well-defined structure at a high level of abstraction is often obscured by troublesome details when you first try to implement it. Many times the confusion is increased because of a lack of description regarding how the high-level ideas are actually implemented. To help reduce the confusion around Truss, this chapter describes the "dance" in more detail.

Overview

This chapter looks at how the "dance" described in the preceding chapter is actually implemented. It shows the order in which each method is called, and describes the files to find the method, or its base. The chapter then looks at the structure for the major components of Truss.

First to be described is `verification_top`, the top-level program of Truss and the base of the "dance." Following this is a description of the methods, and their class, through which files are called for each step.

Then the test component is described. This component follows a dance similar to that of `verification_top`, but for a different set of classes and files.

The `irritator` class is described next. While similar to a `test_component`, irritators have some unique methods worth pointing out.

The last part of the chapter talks about steps that need to be taken to build a new Truss based project, by taking the more-abstract description of classes and applying them to the first few tests in a new project.

About truss_verification_top.sv

When the simulator executes a program, control is passed to the initial block in the `verification_top` program in `verification_top.sv` under the *truss/src* directory. In this handbook we refer to this functionality as the "dance." It is this program block that interacts with your top-level components: the test, the testbench, and the watchdog timer.

Let's look at the dance with respect to the code you have to write. This is illustrated in the figure on the following page. A square box indicates that the method has a default implementation, and a rounded box indicates it needs to be defined for your project.

The Dance - Detailed Flow

The `watchdog` class is already written and should be sufficient for most purposes. (We will not discuss the watchdog timer's methods, because they are relatively straightforward.) You'll have to write the test and testbench classes.

You will have to write a `build_interfaces()` function, along with the `real_interfaces` module that creates the actual chip interfaces. The object returned from the `build_interfaces()` function is given to the testbench constructor.

In the `testbench` constructor, instantiate your generators, checkers, BFMs, and so on. (This assumes that your team has decided to put these objects in the testbench rather than in the test components.) Then add your constraints by using the *dictionary*. These constraints will be picked up by your generators and configuration objects to guide the randomization. Initially, you will probably have no constraints.

The test's constructor will create all the test components and irritators that it needs.

In the `testbench::do_randomize()` method, randomize your local variables and then call `do_randomize()` on lower-level components, as appropriate. Your testbench may have configuration objects for each protocol or feature that is used to configure the chip. You may or may not want to use the built-in SystemVerilog `randomize()` method.

The `test::do_randomize()` method is similar, in that the test randomizes each `test_component` it owns. In addition, the test may select some subset of the components and irritators it owns.

The `testbench::time_zero_setup()` method is where you drive wires prior to letting the chip out of reset. You may need to wait for the PLL to lock, or set up "sensor" pins on the chip in this method.

The `test::time_zero_setup()` method usually just calls all the active test component's `time_zero_setup()`. This is to allow test components that have a "plug-in" behavior, such as USB and PCI Express, to perform their initial training. (To use this method is a judgment call, as you may want to bring up a protocol later in the simulation.)

The `testbench::out_of_reset()` method will bring the chip to a stable state (for example, one that can accept register access). If the team so decides, you could use `test::out_of_reset()` to reset the chip.

The `write_to_hardware()` methods in both the test and the testbench are where you perform register writes to move your selected configuration to the chip. The test's `write_to_hardware()` method usually just calls the same named method on all its test components. This is because the actual register writes will occur in the BFM or driver. One exception is when you are writing a direct test, and it's easier just to write the registers at the test level.

The `testbench::start()` method, if it knows which protocols and features are in use, starts up all the BFMs, monitors, and drivers. Depending on your architecture, it may also start the generators and checkers.

The `test::start()` method usually just calls the `start()` method on all its owned test components.

The `wait_for_completion()` methods in the test and testbench are used to pause the verification system until the test is finished. Although there are many ways to do this, the examples in this handbook just allow the checkers to say when the test is completed.

The `report()` method in the top-level objects is where they report their status. For the testbench, this method is usually appropriate for reporting the configurations selected; for the test, it usually just calls the test components.

That's it. This may seem like a lot of methods to write, but you probably do not need to perform tasks in all the methods. Later in this chapter, we will talk about the order in which you might want to implement these methods.

The Test Component Dance

Did you notice that most of the time the test just called the same named methods on the test component?[1] That's because verification has a fractal structure, with repeated patterns. The top-level dance is repeated, with a few changes, in the test. This time, instead of `verification_top()` calling the steps, the test does. The `test_component` also plays a role,

---

[1] Okay, so maybe remembering back to Chapter 4 is not that easy.

subdividing the `start()` method into several lower-level methods, as shown in the following figure.

Test Component Dance - Detailed Flow

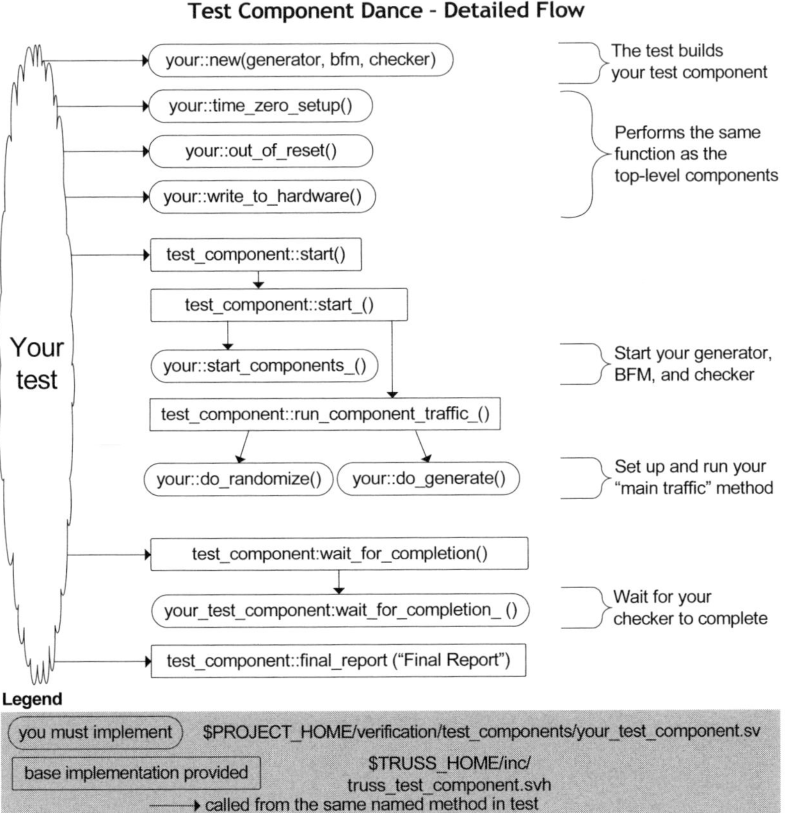

The `run_component_traffic_()` method has a standard implementation, which calls `do_randomize()` and then `do_generate()`. The `do_randomize()` method has the same purpose it had for the top-level components: to randomize your random variables. The next method called, `do_generate()`, picks up the results of the randomization and interacts with the generator to exercise a feature or a protocol of the chip.

Now, it may seem strange that these methods are implemented like this. However, the idea is to separate the various concerns of the test component: starting, randomization, and generation. This, as will be discussed in Part III of this handbook, creates more-adaptable and less-brittle code.

The organization also sets up the *irritator*, making the transition from a fixed test to an irritator relatively painless.

The Irritator Dance

The `irritator` is an inherited class of `test_component`. Its purpose is to generate background "noise" while the test concentrates on some specific area of the chip. In some sense, using irritators is a way to emulate the real world, where many of a chip's features and protocols are used simultaneously.

So what does an `irritator` add to or change from the `test_component`? Only one method is changed, and two methods are added. All these changes involve the new `run_traffic()` method, shown in the figure below.

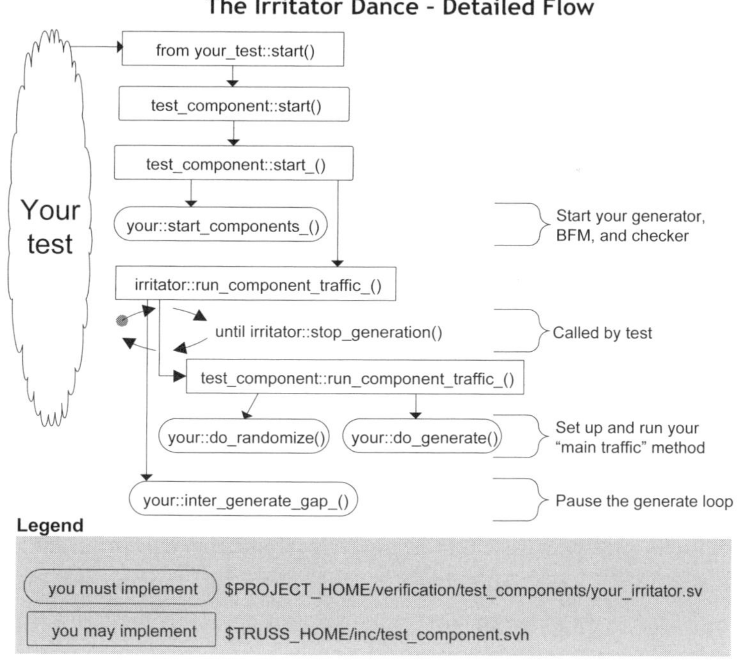

The `irritator` overrides the `run_component_traffic_()` method from the `test_component` base, and calls the base class `run_component_traffic_()` method in a loop. This is the nature of an `irritator`: it just keeps on going until told to stop. The method that stops the loop is `stop_generation()`, which is usually called by your test once the main feature or protocol has finished being tested. This will be shown in detail in the Part IV of this handbook.

One method that you will have to implement is `inter_generate_gap_()`. This method may be empty, for a couple of reasons.

- Your channel has a limited depth, and this limit is used to apply back-pressure to your system.

- Your generator has a built-in delay of some form.

In this handbook we use the checker to throttle the system—because we want to keep a certain amount of data in flight, and the checker is the only agent that knows what has been generated and what has been received. (The chip can handle an unlimited number of back-to-back transactions.)

That's all there is to building an irritator. Note that you will probably start with a test component, and then evolve it into an irritator. It will probably be many weeks into your project before the first irritator is built, but for coverage and finding congestion bugs, irritators are a good choice.

In fact, your first test will probably be even more rudimentary. This first test is the focus of the next section.

Compiling and Running Tests

The sections above described the main building blocks of Truss. The following chapters, as well as later examples, will show how these still somewhat abstract concepts get implemented for real projects. However, before we start looking at more concrete examples, there is one more problem to consider: that of compiling and running a verification environment.

All verification environments need some type of run script to compile and build both the RTL and verification code. In a large project this is not a simple task, because one must track a lot of code, as well as many tools and options.

A goal for Truss is to provide a production-grade run script as open-source components. At the moment, a run script is provided. They are a good starting point for a run script and provide enough functionality to handle the examples in this handbook. It is the authors' hope that through community effort, these scripts can be fleshed out into something better.

The `truss` run script

The `truss` run script controls which files are compiled and run. It is written in Perl and has a number of switches that control its actions. The script will first compile all the SystemVerilog testbench files, and then launch the simulation. After the simulation finishes, the script checks the status of the test run. (This script is used to build and run all the examples that are available at www.trusster.com.) The script is located at $TRUSS_HOME/bin/truss).

Truss uses some environment variables to "understand" its environment. By using environment variables (instead of `.tool_rc` files, for example), the system's assumptions are both obvious and flexible. Truss uses only a small number of environment variables, as listed below.

Variable	Function
SIM	Simulator name (such as ncsim, mti, aldec, or vcs)
SIMULATOR_HOME	Path to the simulator install area
TEAL_HOME	Path to Teal's source files
TRUSS_HOME	Path to Truss install area
PROJECT_HOME	Path to top of the current verification project

The file named `setup` in each of the `bin` subdirectories of each example has default values for the TEAL_HOME, TRUSS_HOME, and PROJECT_HOME environment variables. You'll need to set SIM and SIMULATOR_HOME as appropriate for your environment.

Switches

The `truss` run script has a number of switches to control its execution. Below is a table that expands on descriptions of the most important switches.

Switch	Function
--help	Prints longer help message
--test <test_name>	Runs the $PROJECT_HOME/testcases/<test_name> test.
--clean [options]	Cleans appropriate selection of the system. Default selection is USER. The following options are available: LOGS - Deletes simulation log files HDL - Deletes user-compiled HDL code ALL - Deletes all of the above This switch can be repeated (--clean CPP --clean HDL)
--simulator <SIM>	Selects appropriate simulator from supported list. If switch is not used, then run script reads $SIM. If neither $SIM or --simulator is used script will fail.
--seed <seed value>	Sets random seed to integer <seed value>
--run <number>	Runs the selected test a number of times

For a full description of all switches from a command line, run the following:

```
$TRUSS_HOME/bin/truss --help
```

Using "-f" files

As is customary in the HDL coding world, a file is used to help build the objects. Truss uses a file called `hdl_paths.vc`, which is located in the testbench subdirectory. Almost all of the directories in the examples include an `hdl_paths.vc` file, which lists the files used in that directory.

The `hdl_files.vc` for a basic testbench is shown below.

```
+incdir+$PROJECT_HOME/rtl/uart
$PROJECT_HOME/rtl/uart/uart_transmitter.v
-f $PROJECT_HOME/verification/vip/hdl_paths.vc
-f $PROJECT_HOME/verification/test_components/hdl_paths.vc
$PROJECT_HOME/verification/testbench/top/interfaces.sv
```

The first line adds the `rtl` area to the `include` path. The second line brings in the `rtl` source (there will be more than one of these lines). The next lines direct the compiler to start processing commands from the `vip` and `test_components` area. The last line brings in the real interfaces.

Of course, your `hdl_paths.vc` file will be different, but it probably follows this same format.

The First Test: A Directed Test

Because starting something new is not always easy, this section helps make the process easier by addressing how a first test can be written using Truss. This section concentrates on the steps you need to do, and how a test can be built up from scratch. The next chapter shows a complete first example and focuses more on the flow.

Your first test will probably be a simple directed test, with a `test_component` that does not have a generator and possibly not even a checker. It will probably interact directly with the BFM or driver.

Focus your initial efforts on the driver and the BFM. Write a "first cut" at the driver class, making it have the methods that seem right to you. You may or may not need a monitor, depending on the protocol or feature to be tested.

Define your interfaces to the chip. Then make a module called `real_interfaces` and create the interfaces. Write a class that contains `virtual interface` and build that class in `build_interfaces`.

Next, create a testbench that includes that driver/BFM and think about how to get clocks to the chip and get it out of reset.

Now make a test class and get the whole thing compiling. Before moving on to connecting the test to the driver with a test component, make sure the chip is cleanly out of reset, as this can be done by means of the testbench's `out_of_reset()` method.

The next step is to make a simple `test_component`. This component will probably just be a directed exercise, with perhaps a few reads and writes or just a few calls to the driver. Note that you may use the `test_component`'s pre-implemented methods if you are comfortable with them, but for a first test it might be better just to override the `start()` method directly. This is because that's easier than remembering where to put your randomization and traffic-generation code.

If there is any configuration, use the chip's default configuration. Don't try to randomize anything yet.

Doing the checking can be tricky, so let's worry about that last. We'll probably be looking at waveforms for the first few days anyway.

Now build a test that has your `test_component` as a data member. Initially, have the test call the same named methods on your test component.

Note that the `wait_for_completion()` method probably just returns, if you implemented the `start()` method. However, if you used the `do_generate()` method of the standard `test_component`, you'll want to trigger a `condition` variable at the end of your `do_generate()`. Then, the `wait_for_completion()` would just wait for the signal to be triggered, as shown below.

```
class your_test_component;
  //...your other code here...
   local teal::latch done_;
endclass
```

Then, in the last line of the `your_test_component::do_generate_()` method, do this:

```
task your_test_component::do_generate_();
 //...your directed exercise code here...
 done_.signal();
endtask
```

Then your `wait_for_completion_()` would look like this:

```
task your_test_component::wait_for_completion_();
 done_.pause ();
endtask
```

That's it! You have created your first Truss-based test.

The Second Test:
Adding Channels and Random Parameters

Engineers count "one," "two,"—and then "many." This is because only the first few times they use a technique are significant. After that, everything looks like "many." By writing the first test, we've counted "one." Now we will count "two." The next section will cover the "many."

In this, the second test, we'll get more sophisticated. We'll add the agent layers and also add the generator and checker. These are the steps you need in order to create more advanced, randomized tests. You will probably create several directed tests before you need these additional features, but because this is a book we need to keep moving along.

Remember that the generator and monitor generally have pure virtual methods to communicate the results of their work. We'll add our agents to these methods. There will be an agent for the generator, the driver/ BFM, the monitor, and the checker. Why all this complexity? Because there are many interconnection techniques, each one involving some architectural trade-offs. These trade-offs are talked about at length in the OOP Connections chapter in Part III of this handbook.

To make the connection between the agents, we'll use a Truss channel. So let's digress a bit and look at a channel.

The channel pseudo-templated classes

Verification systems have a lot of producer/consumer relationships. For example, a generator can be considered a producer and a BFM considered a consumer. However, it is a good idea to minimize the knowledge and assumptions of the code interface between these two loosely cooperating objects. One way to decrease the coupling between these components is to use an *intermediary object*. An intermediary object allows the two communicating objects to be anonymous or separated in time. The concept behind this object is called a pipe, mailbox, or channel. Truss uses the term *channel*.

The `channel` class provides the storage for the actual data, as well as the signaling and mutual-exclusion mechanisms. In addition, `channel` also supports a `depth` concept, for designs that want to implement backpressure in that way. The code interface for a `channel` class, as well as the base classes, are in `/truss/inc/truss_channel.svh` in the code that is available at www.trusster.com.

Ideally, the channel class would be parameterized on the type of the object in the channel. However, not all vendors support parameterized classes, so the authors implemented the channel class on an integer type. This is one of those cases where you just have to copy and paste.

The `channel` class also provides for other `channel` objects to be attached to a channel. This allows the data of one `put()` to be replicated across many channels. The common use for this is when a generator creates a data item and both the checker and BFM should get the data. It is also useful if there are multiple listeners to a channel, such as in an Ethernet broadcast, or where there are multiple monitors for a data protocol.

Note that `truss` sends the data to the listeners after the `put` has succeeded for the "main" connection. This is to allow the policies of the main interconnect to throttle the listeners receiving the notification.

Note also that `truss` does not explicitly copy the data. This is because the data in channels of integer types is copied and object pointers are not. In general, the authors have found that it's not necessary to copy the data, as the downstream consumers usually act as readers only.

Building the second test

Now that we have channels, let's use them for the agents. This section is a bit high level, because every situation is different. We'll give general direction, but after you read this chapter, take a look at the next chapter for a first complete example.

Let's say that you are working on a chip interface called `my_interface`. You might have a generator that looks like this:

```
typedef class my_data;
virtual class generator;
    task do_generate(); //make one, then call
                        //done_generate_
    'PURE virtual task done_generate_(my_data d);
endclass
```

We are concerned with the `done_generate_()` method. This is a pure virtual method, so we must implement it in our inherited class. Let's assume we want to add a channel as the connection policy, like so:

```
'include "generator.svh"
'include "my_channel.svh"//cloned code from truss_channel
  class generator_agent extends generator;
    local my_channel out_;
    extern function new (my_channel out);
    task done_generate_(my_data d);
      out_.put(d);
    endtask
  endclass
```

By building a `generator_agent`, we have abstracted how the generator gets the created data to the driver/BFM.

A similar situation exists in the monitor:

```
typedef class results;
virtual class monitor;
extern function new (virtual my_interface mi);
    task start();
    //the connection method
    'PURE virtual void data_received_(results);
endclass
```

And likewise for an agent for the monitor:

```
'include "monitor.svh"
'include "results_channel.svh"
  class monitor_agent extends monitor;
    local results_channel out_;
    extern function new (results_channel out);
    virtual task data_received_(results r);
      out_.put(r);
    endtask
  endclass
```

But what about the other side of the channels? These objects are the `driver_agent` and `checker_agent`, respectively. Their job is to take the data out of a channel and act on the data.

Remember, we are discussing channels here because that's how we wanted to implement the agent layer. This could have easily been a more generic producer/consumer model, or an event-driven method, but implement what feels correct for you. (All the examples in this handbook use channels.)

Here are the classes for the driver and checker and the inherited classes for their agents:

```
class driver;
extern function new (virtual my_interface mi);
    extern task send_data (my_data d);
endclass
'include "data_channel.svh"
class driver_agent extends driver;
  local data_channel drain_;
  extern function new (data_channel drain);
  //must have a start to drain the channel
  task start_();
    fork
      forever begin
        send_data(drain_.get());
        end
      join_none
    endtask
  endclass
  class checker;
      extern task check_data (my_data d, results r);
```

```
endclass
'include "data_channel.svh"
class checker_agent extends checker;
  local data_channel generated_;
  local results_channel checker_in_;
    extern function new (data_channel generated,
                          results_channel checker_in);
    task start();
      //Check the data!
      fork
      forever
        check_data(generated_.get(),
                    checker_in.get());
      join_none
    endtask
endclass
```

The authors realize that there is a lot of code to look at, but just skim it over to get the general idea. The general technique is to inherit a class, add a channel, and append `_agent` to the name.

After the agents have been built, they should be added to the testbench. The testbench holds the generators, drivers, monitors, and so on. The test, on the other hand, holds the test components.

Building the second test's test_component

The `test_component` is relatively straightforward. A `test_component` constructor takes in the parts of the testbench you need. Remember, the entire testbench is not taken as a parameter, because then we would have to make assumptions about the name of the parts we needed. Also, by taking in only the parts we need, several of our test components can be used in the same chip.

The most likely candidates for the constructor's parameters are the generator, the driver, and the checker.

The rest of the test component usually just forwards its calls to the appropriate objects. An example test component is shown below.

```
typedef class bfm;
typedef class generator;
typedef class checker;
```

```
'include "truss.svh"
class a_test_component extends truss::test_component;
   local generator generator_;
   local bfm bfm_;
   local checker checker_;
   extern function new (string n,generator g,
                        bfm b, checker c);
   virtual task time_zero_setup ();
    bfm_.time_zero_setup();
   endtask
   virtual task out_of_reset (reset r);
     bfm_.out_of_reset(r);
   endtask
   virtual task do_randomize (); /* next section */;
   virtual task write_to_hardware ();
     bfm_.write_to_hardware ();
   endtask
   protected virtual task do_generate ();
     generator_.do_generate ();
   endtask
   protected virtual task wait_for_completion_ ();
     checker_.wait_for_completion ();
   endtask
   protected virtual task start_components_ ();
     bfm_.start(); checker_.start ();
   endtask
endclass
```

Although your actual test component will be a bit different from the code above, the general form will probably be the same.

Adjusting the second test's parameters

As soon as you introduce randomization into a test, you'll probably want some knobs to control the randomization. Sweeping most parameters through an entire integer range would chew up a whole lot of simulation time. Besides, it's probably either (1) not interesting, or (2) unacceptable to the register associated with the integer.

A knob is a technique that uses other variables to control the range of a random variable, either directly or indirectly. In this example we'll

 • • • • • • •

concentrate on controlling the random variables directly. (The examples in the handbook use the Teal dictionary feature to pass parameters from a number of sources to the method that will use the knob variables.)

For example, consider a test for a CPU. Assume that a `cpu_generator` class has a `send_one_operation()` method that is called by a `test_component` to tell the `cpu_generator` to create one random operation. The generator is guided by dictionary variables. It is best to put the variables to randomize in a separate function at the top of the source file, because the seeding depends on line number. That way, the sequence of values selected does not change if the code below is reorganized. Of course, new random values chosen will be different for each master seed.

Here is an example function for generating the `operand_a` variable of a CPU operation:

```
class cpu_generator;
  local uint32 min_operand_a;
  local uint32 max_operand_a;
  local function uint32 get_operand_a(uint32 min_v,
                                      uint32 max_v);
    return $urandom (min_v, max_v);
endclass
```

In the `cpu_generator::new()`, the following lines could be used:

```
min_operand_a =
  dictionary::find(name_ + "_min_operand_a", 0);
max_operand_a =
  dictionary::find(name_ + "_max_operand_a", ~0);
```

In the `cpu_generator::do_randomize()`, the following line would be used:

```
operand_a = get_operand_a(min_operand_a, max_operand_a);
```

This same style is used for the other operand and the operator variables.

Now SystemVerilog does have a randomization feature. Later examples will show how you might use them. Be careful, though—randomization tied to the object and its hierarchy can be cumbersome.

So who sets the knobs? There are four ways:

- Use the default specified in the `dictionary_find()` call as the second parameter.

- Put the knob value on the command line.

- Use a knob configuration file.

or

- (Finally) Write code to use the `dictionary_put()` call, which is the mechanism used in our example.

Note that because the Teal dictionary is used, both the command line and the knob file can be added later without the need to modify any of the example code.

The test constrains the test component with respect to the number of times the generator is called. Of course, this specifies the number of operations sent to the arithmetic logic unit (ALU). The code is shown below.

```
teal::dictionary_put(test_component_.name +
    "_min_operations", "4",
      teal::dictionary_default_only);
    teal::dictionary_put(test_component_.name +
    "_max_operations", "10",
    teal::dictionary_default_only);
```

Note that the `name` of the `test_component` is used. This allows the test to pick any name for the `test_component` and still have the code work. It also provides for different parameters for different instances of the `test_component`.

However, be careful with the spelling of the knob variables. They must be spelled the same in both the `find` and the `put` routines in order to make a connection.

Now that the randomization and knobs are connected, we have completed writing the second test. In some ways, this test is rather sophisticated. It uses the Truss framework, and adds agents by using channels to connect the wire-layer classes to the transaction-layer classes.

The testbench created and wired up the generator, driver, monitor, and checker. The testbench can bring the chip out of reset and start the monitor.

The test itself is rather reasonable. It creates and connects the test component to the generator, driver, and checker in the testbench.

The Remaining Tests: Mix-and-Match Test Components

So now what do you do after creating this second, more-sophisticated test? You do what we verification engineers always do—create more tests! As these tests are being written, new test components will also be created, some of which could be used in several tests. Deciding which test components to adapt to different tests is the major activity (besides writing more tests) after you have written the first two tests. This is the "many" count that we talked about earlier.

Of course, you'll be doing other test-related activities, such as adding randomness to the existing tests and looking over your verification test plan to make sure you know when you're done.

And how do you go about adapting a test component from one test into another? You could just put the new test component in the test and wait until both of them are completed. However, as explained in the Truss Basics chapter, there is another way: use the Truss concept of irritators, and warm over, or "recrystallize," the existing test component to an irritator.

Converting the test components to irritators usually just involves deriving the existing test component with the `truss::irritator` component. Then, the appropriate methods will be overridden and the only method you have to write is `inter_generate_gap_ ()`. There are many ways to implement a gap, from the simplest (pausing a number of clock cycles), to the more complex (using back-pressure and bursty traffic). If the checker were inherited from Truss's checker, you can also just wait for generated data to be checked.

This process of writing a new test continues for all the rest of the features and protocols of the chip. Remember, the more irritators a test has, the more likely it is to model what actually happens when the chip design is realized in silicon.

Summary

This chapter tried to clear the fog of how to go about using Truss. We started with a review of the top-level dance, and then showed that the dance also existed in other layers of the system.

We looked at the tools provided by Truss, which is the `truss` execution script.

We covered writing the first test, concluding that it will probably be a directed test. Then, we took the test up a notch, adding connection agents to the generator, driver, monitor, and checker. We introduced the Truss channel as the interconnect technique, but noted that there are many other techniques.

We looked a bit at control knobs, a technique for passing parameters to constrain randomization. (There are many techniques for constraining random-variable generation.) This chapter showed how to harness Teal's dictionary to hook up bounds for randomization.

We finally discussed what to do after the second test. The idea is to write more tests for that protocol or feature, and also test the rest of the chip. The key part of writing more tests is to keep an eye out for what you can "steal" (rather, "adapt") for other tests. By creating irritators, you can use the functionality of other tests as background activities. In this way, the chip is stressed more—and more faults are found prior to production.

Truss Example

*I know that you believe you understand what
you think I said, but I'm not sure you realize
that what you heard is not what I meant.*

Robert McCloskey

Coding is tricky, because we take the great ideas, techniques, and trade-offs and actually make decisions. We put fingers to the keyboard, and decisions are made and trade-offs are fixed in code. Furthermore, learning a new technique only makes the coding task more difficult. An example, or several examples, can help put the technique into perspective.

This chapter is the first example of how to use Teal and Truss in a verification system. It's useful to build and run some example code when learning something new. So download and install the code from www.trusster.com and noodle around with it a bit. You can add print-outs and change the code a bit.

If you want, use this chapter as a guide to some of the more interesting parts of the code examples provided. This chapter is not quite a map to the "homes of the movie stars." Instead, it is more like a mariner's map. It helps you navigate in tricky waters.

Overview

This chapter provides a first complete example of using Truss, where you can actually compile and run the code. The code is not as complex as what you would encounter in a fully featured chip. However, all the main parts are here to consider. The source files may seem silly or overly complex for the chip we are trying to test, but we are trying to demonstrate how to structure a verification system for a real project. Your chips will have plenty of complexity to manage.

This chapter does not walk through every code file. We are all capable of reading code. What it does instead is look at some of the more important aspects of the verification system.

Directory Structure

In order to help you navigate the source files, it's good to show the main directories that comprise a Truss-based system (shown below). We've also included only the main files we will be working with.

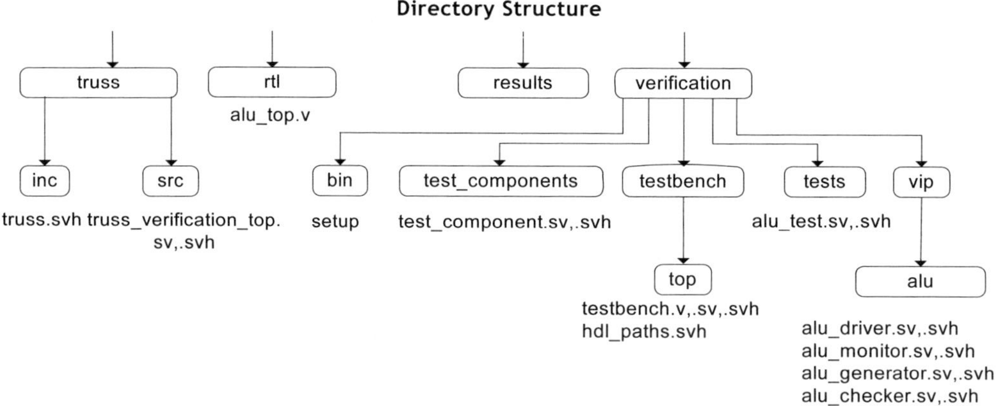

The source code for the chip is in the `/rtl` directory. How does the `truss` run script know this? The file `/verification/testbench/top/hdl_paths.vc` is used to specify the paths to the RTL and the RTL include directories and all source files. This is so that the RTL files can be rooted in a place different from the verification directory.

The `/results` directory is where you run the tests from. It also can be wherever you want. The authors generally put this directory in some non-backed-up networked storage area that is independent of the source-code control system. In the handbook example, the `/results` directory is placed in `/examples/alu`, at the same level as the `/verification` directory.

The `/verification` directory contains all the source code for the verification system. The `/bin` directory is there for the project's local scripts. The authors usually put a setup script there and alias `setup` to it.

The other four subdirectories—
`/tests, /testbench, /vip, and /test_components`
—are where the actual source files are. The `/tests` directory is where your `test_name.sv` and `test_name.svh` exist. These files are used when you give the `--test <test_name>` option to the `truss` script.

The `/testbench/top` directory contains the SystemVerilog and HDL sources for the top-level testbench. If you have more than one chip in your simulation, it may be useful to have `/testbench/<chip_name>` directories.[1] The chapter on Chip-Level Testing shows an example of this.

The `/vip` directory is where chip protocol classes go. There should be a subdirectory for each protocol and major feature you need to test. The idea is that the code in these directories is fairly portable, and may contain purchased VIP as well as project and company-created VIP. In our example, there is only the `/alu` directory.

The `/test_components` directory contains the scenarios that you want to run. For this example, we'll run only one scenario, called `test_component`.

---

[1] Use the --config option to the `truss` script to select a directory.

Theory of Operation

We'll be testing a really basic ALU chip. It takes in two 32-bit operands and performs a simple logic or arithmetic function. We'll use a legacy c-model for comparison with what the chip produces. The output consists of a result and an "operation complete" status interrupt. The `test-bench.v` will instantiate this ALU module and provide system clocks for the chip and verification system.

The main objects are shown below.

ALU Example: Objects and Connections

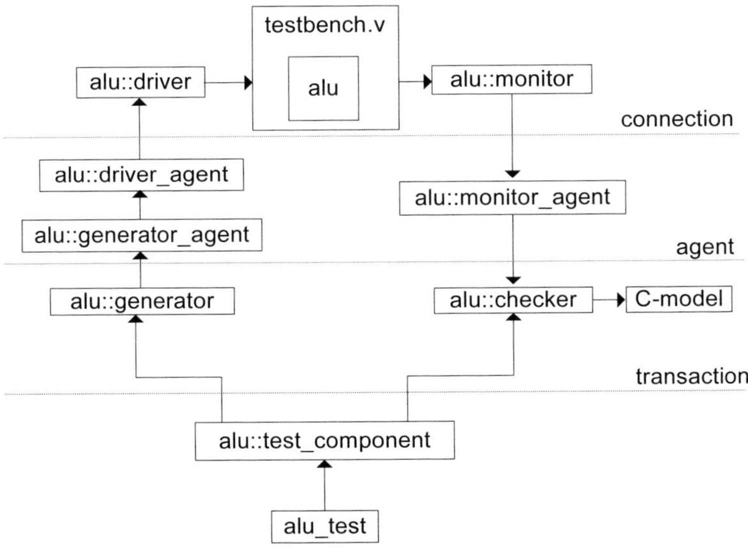

Because there is only one protocol in the chip, we'll just refer to the components by their functionality. In other words, we'll say "driver," although in a chip with many drivers we would need to say which protocol we are talking about. (Note that we do prefix the code in an ALU string.[1])

[1.] The authors had intended to use a package, but the SV language is fairly weak with respect to packages. For one thing, one cannot extend a package, which meant all the header files would have to be included from one package/ endpackage declaration.

In the testbench class, we have all the classes of the ALU component layer. There is a connection-layer driver and monitor, with their accompanying agents. There is a generator and a checker. The checker is interesting, because we have a legacy c-model of the chip, which will be used by the checker.

There is also a `test_component` class, which runs a random number of operations through the chip. And, of, course, there is an `alu_test` class, which builds a `test_component`, giving it the generator, checker, and driver from the testbench.

The following illustrates the wires used by the verification system:

ALU Example: HDL Connections

The driver and the monitor take care of the protocol into and out of the chip. The testbench takes care of bringing the chip out of reset.

The remaining sections highlight some specific "points of interest" in the code. The code itself, being the first example, is not that big. If you want to follow the code through its execution, start with the Truss `verification_top.sv`, then move on to `testbench.sv` and `test.sv`.

Running the Simple ALU Example

You might want to see the log messages on the screen, so let's talk about how to run the example. In the `/examples/alu_tutorial/bin` directory, there is a setup script. If you look at the `setup` file, it sets up a few environment variables that are needed by the `run` tool.

First, source the `setup` file, then execute the following:

```
$TRUSS_HOME/bin/truss --test tutorial_test
```

The `truss` command has many more options; type `truss --help` for a synopsis.

You should see the source files being compiled, and then the test should run. When the test runs, a series of printouts will announce the flow through the test.

Points of Interest

The next few sections address specific places in the code. These sections follow the general way you go about hooking up a chip to a Truss-based verification system.

For example, the first thing to be concerned with is bringing the entire chip out of reset. After that, you'll probably want to pick a chip protocol and write the driver and monitor classes for it. Then, you might decide upon some specific operations you want to perform and write the test component to exercise the protocol or feature.

In general, the test builds the test components and ends when the last operation completes—that is, when the test component's `wait_for_completion()` returns.

Power-on Reset

Most chips have a power-on reset sequence. This sequence can be basic, or rather complicated. In this example we address a basic sequence.

The chip has a *reset* line, which is pulled low to initiate a reset. After the line is asserted, the chip performs its reset sequence. This chip only needs a fixed-duration pulse.

The `testbench` class is responsible for bringing the chip out of reset. The testbench methods `time_zero_setup()` and `out_of_reset()` are called by the top program to reset and configure the chip. In our ALU example, we'll use a reference clock to count a number of cycles to keep the `reset_n` low.

Below are the snippets of code that perform the chip reset. The methods are located in `testbench.sv`.

This method is called first by `verification_top()`:

```
task testbench::time_zero_setup();
  top_reset_.resetr = 0;
endtask
```

Note that the `top_reset_` was built by the `build_interfaces()` function and then cached in the testbench's constructor.

Then, this method is called:

```
parameter int reset_count = 10;
task testbench::out_of_reset (reset r);
  top_reset_.resetr = 1;
  for (int i(0); i < reset_count; ++i) begin
    @ (posedge (top_reset_.clock));
  end
  top_reset_.resetr = 1;
endtask
```

That's all there is to it. Now the chip is ready for operation.

Driver and Monitor Protocol

Now that the chip is out of reset, we can start to drive it. This chip has a simple protocol for sending operations to perform. Assuming op_done is asserted, the driver puts op_code, operand_a, and operand_b on the wire. Then it asserts do_op and waits for op_done to be asserted. The code to do this is in alu_driver.sv and is shown below:

```
task alu_driver::send_operation(operation op);
  alu_input_.op_code <= op.op_code;
  alu_input_.operand_a <= op.operand_a;
  alu_input_.operand_b <= op.operand_b;
  alu_input_.op_valid <= 1;
  //Now wait until accepted
  @ (posedge (alu_input_.operation_done));
  alu_input_.op_valid <= 1;
  @ (negedge (alu_input_.operation_done));
endtask
```

The alu_input_ above is a virtual interface to an ALU interface, which was passed in to the constructor. Note that in a "real" driver, you might want to put #(drive_delay) before the first assignment.[1]

The monitor code is fairly simple as well. The monitor uses a local utility class called run_loop. It consists of two methods, loop_condition_() and loop_body_(), which are run in a thread. The idea is that a number of monitors are just infinite loops of "wait for trigger" and then "gather data." This class represents that concept.

The loop_condition_() method of the monitor waits for op_done to go high. The loop_body_() method then copies the result into a local variable. It then calls the pure virtual method receive_completed() to connect to the monitor agent.

[1.] Recall that the authors do not believe it's a good idea to use clocking blocks. You and your team may, however, want to use them.

Here is the code, in `cpu_monitor.sv`:

```
task alu_monitor::loop_condition_();
  @ (posedge (alu_output_.operation_done));
endtask
task alu_monitor::loop_body_(output bit go_on);
  receive_completed_(result_.to_int());
  @ (negedge (alu_output_.operation_done));
  go_on = 1; //continue loop
endtask
```

Other than the reset logic (and the watchdog timer), the monitor and driver are the only code to interact with the chip wires.

Next we'll look at how we come up with the operations to be sent to the driver.

The alu_test_component

We now run a random sequence of operations through the ALU, testing the basic operations with random operands. The `start_components_()` method is used to run this exercise.

The code is shown below.

```
task alu_test_component::start_components_();
  driver_.start();
  checker_.start();
endtask
```

Like most test components, this one just starts the lower-level components.

The `start_components_()` method is used to do select the number of operations to perform.

```
task alu_test_component::start_components_();
  bit [7:0]¹ min_words =
    dictionary_find_integer ({name_, "_min_ops"}, 10);
  bit [7:0] max_words =
```

[1.] Be careful about using byte, as it is signed.

```
            dictionary_find_integer ({name_, "_max_ops"}, 15);
        number_of_operations_ =
                get_number_of_operations (min_words, max_words);
    endtask
    function bit [7:0] get_number_of_operations
                                (bit [7:0] min_v, bit [7:0] max_v);
        bit [7:0] returned;
        'RAND_RANGE (returned, min_v, max_v)
        return returned;
    endfunction
```

Checking the Chip

Because we do verification for a living, the automated checking of the chip's results is important. In our case, we have a legacy c-model of the ALU and will use it to check that the answer is what we expected. The checker waits for the monitor agent to deliver a completed operation. Then it uses the inputs sent by the generator to have the c-model come up with the expected result.

The c-model prototype is shown below.

```
#if defined(__cplusplus)
  extern "C" {
#endif
  unsigned int alu_model(unsigned int a,unsigned int b,
                         unsigned char op);
#if defined(__cplusplus)
  }
#endif
```

Note that the ifdefs allow the code to be compiled by both C and C++ code.

This key algorithm is in checker.sv and is shown below.

```
task alu_checker::start_();
  forever begin
    operation gen;
    teal::uint32 actual;
    generated_.get(gen);
```

  Hardware Verification with SystemVerilog

```
        actual_.get(actual);
        if (alu_model (gen.operand_a, gen.operand_b,
                    gen.op_code) == actual) begin
          log_.info ($psprintf (" EXPECTED %s == sent %d"
                            gen.sreport(), actual));
        end
        else begin
          log_.error ($psprintf (" EXPECTED %s != sent %d"
                            gen.sreport(), actual));
        end
        int count_; generated_.count(count_);
        if (!count_) begin
          completed_flag_.signal();
          return;
        end
      end
    endtask
```

The checker works fine as long as the `operation_done` is in synch with
the result. However, the checker can be wrong if the monitor misses a
result or somehow inserts an extra one. We could have registered the chip
inputs at the same time as we got the results. However, by doing this we
make the assumption that there are no queuing or pipe stages in the ALU.
This assumption works fine for our example, but it is probably not valid
for most ALUs.

Completing the Test

When does the test stop? When `verification_top()` calls the test's
`wait_for_completion()`, which in turn calls the test component's
`wait_for_completion()`.

In turn, the test component's `wait_for_completion()` calls the
checker's `wait_for_completion()`. The authors agree that this sounds
silly, but in the later examples we actually do a bit more than just forward
the call.

In the end of the forwarding chain, it's the checker that actually decides
when the test is done. This makes sense, because the checker is best able
to "judge" what the chip did and when all the inputs have been checked.

But how does the checker know? There are many possible ways, but in this example the checker assumes that when the generated data channel runs dry, the test is over. This is a valid assumption—as long as you make sure that the generator can always be one step ahead of the checker. (If your chip has any latency, this is not a hard assumption to sustain.[1])

The checker code is shown below—

```
task checker::wait_for_completion();
  completed_flag_.pause();
  //note that the checking thread completed normally
  completed_ = 1;
endtask
```

—and at the bottom of the main check loop:

```
    int count_; generated_.count(count_);
    if (!count)
      begin
        completed_flag_.signal();
        return;
      end
```

Remember that after the `wait_for_completion()` returns, the top calls the `report()` method in the test. The test calls the test_component's `report()` method, which in turn calls the checker's `report()` method.

The `report()` method prints the state of the `completed_` boolean. In this way, when you have multiple test components and the watchdog timer shuts the simulation down, you can tell which checkers have not completed.

---

[1] Note that an intergenerate delay should not affect when the expected data are sent to the checker. The point is that even when delays are inserted, this model should be valid.

Summary

This chapter is a tutorial on the Truss framework. We exercised a simple ALU, but implemented all the parts of a Truss-based verification system. The main objects and their connections were shown. The directory structure was introduced so we can find our way around the code. Then, the chip and the HDL connections were shown.

After laying out the verification system and showing how to run the example, we looked at how the chip was to be brought out of reset. We did a quick side tour to talk about how to run the example. Running the example produces many log messages, but this is probably a good thing when one is learning.

We showed how to bring the chip out of reset and how the driver and monitor connect with the chip. One point to note is that while this protocol required only a few wires, many real protocols are no more complicated. Of course, your code will be more detailed.

We looked at an important part of the verification system, the checker. In this example, the checker used a c-model to check that the chip was working correctly.

The last thing we looked at was how the test stopped. We looked at the normal path, ignoring the watchdog timer. We showed how the checker was in charge, pausing the end of the test until all the data had been checked. The interesting point to note is that the checker may have had errors, but it will continue until all generated data have been checked. The Truss utility class `error_threshold` can be used to terminate the simulation in the case of excessive errors. The Truss `verification_top()` also does this.

Whew! We made it through the first example. Time for a coffee break and some foosball!

Part III: Using OOP for Verification (Best Practices)

This part of the handbook explores what it means to write OOP-based code. It's not easy to "get it" when it comes to OOP. There are many techniques, and experience plays an important part.

We'll walk through the activities of programming, showing examples and experiences that form the design and coding biases often found in OOP-based verification systems.

We'll end each section with a short sentence about the lesson learned from each example or experience. This is in no way meant to be a rule. Rather, it's another trick, to be added to your bag of tricks you can use— or not—as appropriate.

This part addresses the following themes:

- The shift in thinking that usually occurs when you start working with OOP

- How to use OOP to manage complexity when architecting a verification system

- Techniques useful in making classes and connecting them
- Code techniques useful in writing OOP-based code

Thinking OOP

C H A P T E R 9

Monty Python, episode 15, 1970

Getting your brain around OOP is a challenge. You may have followed the syntax of classes, inheritance, and so on. But when should you write new classes or use inheritance? What about owning an instance versus deriving from a class? A little befuddlement is okay—OOP requires a shift in thinking, and mental fog is a natural result.

This chapter will get you "thinking OOP." A reason OOP is all muddy is that there are no rules. "Thinking OOP" is more about using a set of coding biases and lessons learned than in making trade-offs. Sure, we could have pretended there were rules, providing numbered steps such as, "first you must blah, blah, blah," or "you must always apply by blah, blah, blah," but no one would remember. Instead, this handbook tries to teach you how to ride the "OOP bicycle." Learning to "think OOP" is not trivial, but once you've learned, you never forget.

Overview

We now introduce thinking and using OOP in stages. From the first stage (the "big picture") to the last (coding), we introduce an "arsenal of weaponry" that has proved useful for programmers. This arsenal requires a few chapters. In this chapter we concentrate on framing the OOP process. We talk about the difficulties in managing complexity and creating adaptable code. We then discuss the difference between the interface and the implementation of a piece of code. Subsequent chapters cover architecture and coding.

Remember, verification is neither simple nor easy. Any serious attempt to verify hardware will result in a complex system. Consequently, it is important to realize that the complexity of a verification system is not the result of poor implementation, but is largely intrinsic to the problem of verifying a complex design.

Object-oriented programming is used to help manage complex problems, not eliminate them. The goal is to make the complex appear simple without introducing unexpected behavior. The trick is to keep the focus on making things seem as simple and clear as possible, while minimizing the use of "magic" code or confusing connections. This will nonetheless create a bit of a conundrum, as what is simple and clean to one is often perceived as unnecessarily complex and "sneaky" by another.

There is no "silver bullet" to slay the werewolf of complexity. Verification complexity needs to be managed differently across different types of projects. For example, System-on-a-Chip (SoC) designs are complex because they often involve several independent input/output (I/O) subsystems. For that matter, advanced processors, especially those dedicated to graphics and audio applications, require multistage pipelines and interrelated computation. The best solution to this complexity is communication, through understandable design and code (abstractions, minimal assumptions, and so on), combined with a drive toward common-sense simplicity.

Furthermore, making the design and code adaptable adds to the difficulty of verification—yet it is exactly this additional difficulty that OOP was created to manage. In a fast-paced and increasingly complex product cycle, writing adaptable code is as important as managing complexity.

The concept of creating and adapting code seems simple enough, although in practice it is very difficult, for a number of reasons. This chapter looks at some of the reasons why building adaptable code is difficult. Don't get discouraged; often adaptable code is a natural by-product of a well-reasoned design.

One way to think about managing complexity and creating adaptable code is to look at what is holding us back. In the real world of verification development, there are rarely perfect solutions. Nevertheless, we can build systems that make appropriate trade-offs. To this end, some sections of this chapter include a table of trade-offs to help you make the most appropriate choice for your code.

Sources of Complexity

When you sit down to write code, there are several constraints that slow down the coding process. These constraints can be viewed as adding complexity, because they make an inherently difficult problem even harder. Some of these complexities arise from external sources such as teamwork (that is, local personalities or working with remote sites). We'll touch on teamwork lightly, following our discussion of complexity.

Other complexities are created when a solution is implemented. This is because any solution, almost by definition, involves trade-offs. The authors call this *implementation complexity*, and discuss it in the next section.

Essential complexity vs. implementation complexity

In any verification task there are algorithms and procedures that are required by the specification. In USB, for example, there is a process called enumeration that has a prescribed algorithm for both the host and the device. This is called *essential complexity*, because it is required. When that protocol is implemented in classes and code, some additional complexity is created. For example, the host and device interrupt code must try different scenarios. The authors call these classes and code *implementation complexity.* The classes and code are needed, but are more an artifact of the solution than a real part of the problem. Why is this distinction important? Because you cannot get rid of the essential complexity, the goal is to make the essential complexity as simple as possible, and keep the implementation complexity as minimal as possible.

Implementation complexity is to some degree always created when you are designing or coding essential complexity. For example, although the PCI Express protocol specifies endpoints and a root complex (the host node, or top of the tree), no data structures are specified to manage these concepts. When these are coded in the verification of a root or endpoint, they are implementation complexity.

Remember, engineering is all about building the appropriate solution to a problem, creating problems as a result of that solution, solving those problems, and so on. The successful engineer transforms the big problems into a series of solutions and little problems that are acceptable for the task at hand.

It is important, as much as possible, to use the terms and connections identified in a protocol, chip, or system specification. This will minimize the implementation complexity and provide a basis for a mental model of operation. Try to minimize the implementation complexity, but understand that it will always be present.

> *Be aware of the essential complexity of the problem, and be even more aware of the complexity created by the solution.*

Flexibility vs. complexity

To make a verification system that is flexible also appear simple is exceedingly difficult. Flexibility and complexity are often trade-offs, and usually flexibility wins. It often helps to keep in mind that the developers who will read your code are intelligent, but time-limited. An overly complex solution will do more to slow them down than a simple, but tedious, interface.

As an example, consider a memory subsystem of a verification testbench. Assume that this memory subsystem is on the main bus of a chip. There are many possible questions to ask when designing the interface. For example, should there be separate back- and front-door accesses? Is randomization needed? Should all writes be checked to confirm that the chip has accepted the data? What happens when the chip reads memory that was not initialized? Is this an error, or should it be ignored (and random or undefined data returned)?

The following is an example of a possible class interface. How obvious is it that the design questions above were answered in a flexible, but not complex, way?

```systemverilog
'include "teal.svh"
import teal::*;
class memory_bus;
    extern function new ();
    //The zero-time memory access methods:
    extern function uint32 back_door_read (uint64 address);
    extern task back_door_write (uint64 address,
        uint32 value);
    extern function uint32 front_door_read
        (uint64 address);
    extern task front_door_write (uint64 address,
        uint32 value);
    //Will randomly select front or back door every time
    function uint32 read (uint64 address);
      bit front_door; 'RAND_RANGE(front_door, 0,1);
      return (front_door ?
              front_door_read (address) :
              back_door_read (address));
    endfunction
    task write (uint64 address, uint32 value);
      bit front_door; 'RAND_RANGE (front_door, 0,1);
      if(front_door) front_door_write (address, value);
      else back_door_write (address, value);
    endtask
    'PURE virtual uint32 handle_DUT_unitialized_read
                        (uint64 address);
    endclass
```

There is no immediate solution to the flexibility-vs.-complexity trade-off. The "current best" answer will evolve as your team changes its members and gains experience. The class above certainly seems complete, if possibly a bit too complex. One thing to note is that there is both front- and back-door access as well as a random method. This seems overly complex, as the random method could be implemented in an inherited class if that is what coders want. In this case, that interface should probably be removed from this base class.

Now suppose that team members designed their code to work independently of whether the memory read/write method was front or back door. In this case the random method should be the only approach, and the front- and back-door accesses could either be moved into the private access or left to subclasses to implement. Note that by removing the explicit calls to front- and back-door access, we are making the code both less clear and more flexible. This is either good or bad, depending on whether the team wants to write code that is independent of the front- and back-door access method.

Now take a look at the pure virtual method to handle a read to uninitialized memory, `handle_DUT_unitialized_read()`. By making this method pure, an inherited class must be created. However, even this is confusing. Is this the method for a verification-initiated read or a chip read? Consequently, there should be two methods to cover both cases. Also, while a flexible solution requires two methods and an inherited class, it may be appropriate to make a simplifying assumption.

Suppose that the team considered a read to uninitialized memory by the verification system to be an error. This could simply be written into the implementation of the read method. However, the chip side is not so clear, so maybe just returning X's might be the team's preference, and this pure virtual method could possibly be removed.

Here is an abbreviated memory class resulting from the previous discussion:

```systemverilog
'include "teal.svh"
import teal::*;
class memory_bus extends verification_component;
  extern function new ();
  //The zero-time memory access methods:
  extern function uint32 back_door_read (uint64 address);
  extern task back_door_write (uint64 address, uint32
      value);
  extern function uint32 read (uint64 address);
  extern task write (uint64 address, uint32 value);
endclass
```

*A class interface can be flexible or simple, depending on the
specific needs of the verification effort.*

Apparent simplicity vs.
hiding inherent complexity

One of the goals of good coding is that there should be no surprises when
one tries to understand the code. As a counter-example, an interface may
appear simple, but in practice it may have a usage model that affects the
simplicity of the interface. This often shows up when you try to inherit
from a class or call the methods in a different, but rational, order compared
to what the original coder intended.

Example: How hiding complexity
can create confusion

Here is an example of where "hiding the complexity" actually made the
system harder to understand. In verification there are classes that manage
the top level of a subsystem, other classes that manage the transmission
of data (often called stimulus generators), and still other classes that
monitor the output of the chip. One verification team decided to put these
concepts into separate base classes that all used a common root class.
The common root class had the usual `init()`, `start()`, `stop()`, and
`post_run()` methods. All of the subsystem classes inherited from these
base classes. The constructor of the base class maintained a master list
of all the instances, with an `enum`, called `type_id`, to indicate the type.
Then, when the verification system started up, the "program" base class
would walk this master list and call the `init()` methods of all the "top"
(that is, `type_id==top` objects first, followed by the `init()` method
of all the "monitor" objects, in turn followed by the `init()` method of
all the "generator" objects, and so on. The actual system had ten different
flavors of `type_id`, and thus ten different passes for each method.

Not surprisingly, this "under-the-covers" magic caused significant difficulties for the team. It was hard for subsystems to control which
monitors got started in what order, except by carefully controlling which
objects were constructed when. Engineers new to the project would get
confused and fail to understand the hidden priorities. The team tried to
solve the problem by adding a special `StartupClass`, which would be

the first to run its `init()` method. However, this made the effort of moving a test from the unit level to the full chip level difficult, because the single `StartupClass` could not be reused. As a result, the "simple" system ended up adding substantially to the complexity of the verification effort.[1]

Example: How apparent simplicity leads to later problems

Here is another horror story. Almost every chip has an interrupt capability. In one case our test team decided to have a single interrupt scoreboard for a chip. The scoreboard would not check the reason for the interrupt; instead, it would simply have a queue of interrupt handlers and make sure there were not any unexpected interrupts or leftover handlers. In practice, this simple scoreboard turned out to be inappropriate for several of the major sources of the interrupts.

There were two main classes in this case. The first was the interrupt handler, which was used to encapsulate the handler logic if it matched the interrupt. This class is shown below.

```
class interrupt_handler extends verification_component;
    extern virtual function bit match_id
      (uint32 vector_id);
    extern virtual task do_handler ();
endclass
```

The next class was the interrupt scoreboard. This class had a list of handlers as well as a `start()` method to watch for interrupts. It also had a `post_run()` method to make sure there were no unused interrupts. This class is shown below.

```
class interrupt_scoreboard;
    extern task post_handler (interrupt_handler ih);
    extern virtual task start ();
    extern virtual task post_run();
endclass
```

[1.] One could reasonably argue that this is just an example of a poor or inappropriate design, yet the authors have seen it used in two different companies.

 • • • • • • •

When an interrupt was asserted, the scoreboard called `match_id()` for every `interrupt_handler` on its scoreboard. The first interrupt for which `match_id()` returned true would be removed from the scoreboard, and its `do_handler()` would be called. It was up to the test writers to be as specific as they wanted to be in the `match_id()` method. Some test writers always returned true if the interrupt was for them, whereas other test writers tried to be more specific as to the exact reason for the interrupt.

This class worked fine until the team started testing chip interfaces for which one could not reasonably predict the number of interrupts. Two interfaces in the chip had this property—the USB and the Ethernet subsystems.

In the USB subsystem, a start-of-frame interrupt was generated once every millisecond. Because the USB controller initialized its start-of-frame counter to a random value, and the test end conditions were fuzzy (they were simply based on other data streams draining their checkers), the number of start-of-frame interrupts could not easily be predicted (nor was this number very interesting to know). In this case, the original decision regarding where to put the scoreboard caused an almost impossible checking algorithm for the USB subsystem.

The other example from this same chip was related to the Ethernet unit. The generator randomly assigned masks and packet types, so predicting where (or whether) a packet would arrive was difficult enough—let alone predicting the interrupts that would be generated.

The final straw was that, as an optimization, the chip combined interrupt events, so that if two interrupt-generating events on the same subsystem occurred before these events were serviced, only one interrupt would be generated. Accounting for all these possibilities was not only very hard, but it was also was of little use for verification. As a result, the interrupt scoreboard was removed and individual subsystems were called to handle all interrupts, based on a fixed vector identifier-to-subsystem mapping.

Although a resulting class interface might be too simple for what must be accomplished—and overly simplistic models can lead to complications in implementation—don't stop striving for the simplest usage model possible.

Team dynamics

It may not be obvious, but the makeup and operation of the team affect not only how code is created, but also how well the adaptable code is received. This, in turn, affects the success of the project. Why is this relevant to "thinking OOP?" The addition of OOP created a much more tightly coupled architecture—one where understanding the intent of your fellow team members is essential to coding well. OOP is likely to bring into focus any team issues already present.

A healthy team is better able to create well-built code and adapt existing code. What does this have to do with a handbook on verification? Verification systems have become as complex as production software. As a result, team dynamics becomes a major factor in the success of the verification effort. The sooner we, as an industry, realize this, the sooner we can address team dynamics. *Team dynamics is the current focus of the software domain.* (There are books on this in the For Further Reading section at the end of this chapter.) As this is a new concept for verification teams, we'll just touch on the subject here.

Team roles

There are many team roles and responsibilities. A clear mapping of roles to personnel is necessary for a well-functioning team. An important role is that of the *code leader*. The person in this role is considered the "godfather" of the team and usually is consulted on major and minor architectural decisions. This person knows the language thoroughly and is interested in the latest "best practices." Another important role is that of the *technical leader*, who knows not only the architecture, but also the scripts, policies, and assignments of the team. This person is different from the code leader in that the responsibility of the technical leader is broader, with a more project-level view. The role of *toolsmith* is also critical. This person provides all the scripts and "spells" that make the day-to-day learning and writing of the code easier.

Identify and celebrate team roles—they are all equally important to the success of the team.

Using a "code buddy"

Often an independent reviewer can find places where the code can be made clearer and more adaptable, because the reviewer can concentrate on the finished product, not on the failed attempts. Selecting the right code reviewer is critical to the success of this endeavor.

It is almost always a mistake to have a team code review. This creates a poisonous many-against-one atmosphere. Instead, let each coder pick their personal "code buddy." This individual will be a trusted coworker with whom an informal walk-through of the code can be accomplished. The focus should be on the fact that the code review happened at all, not on the specifics of the review.

Code reviews, though necessary, must be done with care. Otherwise, team cohesion and the project will suffer.

Creating Adaptable Code

How is a section on creating adaptable code relevant to "thinking OOP?" Well, one of the reasons OOP has proved useful is that it really helps create code that is adaptable. In some sense, "thinking OOP" is about creating adaptable and reasonable code.

The term *code adaptability* is an informal measure of how easy it is to move code from one use to another. Code adaptability is an acknowledgment that there is more work to do, even if you purchase verification IP.

Achieving adaptability

Achieving adaptability is a fancy way of saying that you and your programming team creates code that has proved useful for generations of chips. Of course, there is always new code to write for each chip; otherwise, there is little reason to create another chip. The idea is to build adaptable code that is acceptable in its complexity, can be reasoned about, has minimal assumptions about its environment, and has minimal connections to other code modules, *as appropriate for your organization.*

As an example of code adaptation, one company may prefer to make a copy of some common code before starting a project. In this way they can remove unnecessary code and complexity and, at the corporate level, handle the management activities of common bug fixes. Another company may prefer to keep a single code base for all projects. This will almost definitely increase the code's complexity and size, yet common bug fixes are automatic. Each approach is justified; as is a common theme in this handbook, there are no absolute correct answers.

Let's come up with some of the ways code can be adapted. You can do the following:

- Reuse existing code without any modifications, or
- Copy existing code, or
- Use existing code as base classes, or
- Use only the test cases of existing code, or
- Use only the BFMs of existing code

Here, the meaning of "existing code" includes both in-house and purchased Verification Intellectual Property (VIP).

> *The premise of creating adaptable code is that, for the next project or revision, it will be faster to adapt existing code than to develop it anew. This is a major reason for using OOP techniques.*

Why is adaptability tricky?

So most people want their code to live forever. In practice, however, creating adaptable code is difficult.

One tricky thing with adapting code is that the definition of "adaptable" is relative. Does this mean the code can be compiled on different versions of a compiler? Does this mean the code can be reused in a different project? What about VIP, which can be adapted to different compilers in different projects in different companies? As with the definition of verification, each organization will have different metrics for what makes code "adaptable." The requirements will often change, depending on the team, the individual coder, and the purpose of the code.

While the need to create adaptable code is always present, the actual use of adapted code evolves as both a team and our industry gains experience in reuse techniques. The growing interest in using OOP is a good example of this evolution.

Another difficulty with adapting code is that production code is ugly. A verification system that has made it to tapeout is often riddled with workarounds and undocumented assumptions. It also contains features specific to a particular chip. Moving the code to another project is not trivial.

There are more barriers to adapting code, such as a heterogeneous team experience, a lack of domain experience, and competing requirements for code flexibility.

Creating adaptable code may appear difficult at first glance. However, with the techniques presented in this handbook, you can produce code that is adaptable to a large range of projects, with a minimal increase in the code's complexity.

Architectural Considerations to Maximize Adaptability

This section looks at the reasoning behind, and techniques for, recognizing and building adaptable code. (The next chapter looks more in depth at these considerations). To create an adaptable architecture, begin by asking the following fundamental questions:

- Where and what are the global components (such as a memory map)?

- Where do the lines of responsibility lie?

- What capabilities are needed in the current—and the next—design?

A factor affecting the ease of code adaptation is how close the next chip is to the current one. Obviously, the closer the two chips are, the better the opportunity for adaptation. This touches on the subject of minimizing assumptions in a functional area. The fewer the chip-specific assumptions, the more likely the code can be adapted. However, the resulting code may be more complicated and take more time to develop and learn.

Building adaptable code also involves a *correct by construction* technique (discussed in detail in the next chapter.) This means that, once you change the code for the next design, it is better to have the code fail to compile, rather than having it fall off an "if" test or just return. If the code does not compile, there is obvious work to be done. The worst case is code that compiles and runs, but is wrong. Code assertions, such as the `assert()` function, can also help to catch the runtime errors. In other words, use assertions for assumptions that cannot be expressed by construction.

Changes are easy—or just plain impossible

An interesting phenomenon occurred when large-scale systems started to be built by means of OOP techniques. The developers found that changes were either quite easy or nearly impossible. They were easy if the architecture anticipated the change, which usually meant the change occurred along class lines. On the other hand, changes were really, really difficult if several items of data to be changed were spread out into multiple classes. Of course, this is code, and code can always be made to "work," but the very technique of hiding data and making data (objects) drive the algorithms means that some algorithms may be spread out across many objects.

This is relevant because it means that if a feature is not present in the original code, it may be difficult to add that feature later. On the other hand, it might be simple. This leads to a design bias that favors smaller classes, with attention to the assumptions made in those classes. A few examples illustrate this point.

Consider an object that needs to access external RAM as part of its function. If the current design used ZBT[1] RAM, a certain protocol is used. If, in a later design, the RAM were changed to QDR[2] RAM, a new

---

[1] Zero-bus turnaround.
[2] Quad data rate.

protocol would be used, but the essential function would remain the same. If the original designer had "hidden" the actual memory interface in another object, then it would be relatively simple to build a QDR object and pass that into the other object's constructor.

As another example, consider a testbench object that supports the chip in its initial boot sequence. Suppose that a chip implementation supported booting from a UART or USB. In this case you could code the different protocols in the testbench directly, or implement them as specific inherited classes of a generic `boot_source` class. Suppose, also, that later implementations of the chip might support additional boot devices, such as a disk or I^{2}C interface.[1] The test could then let a testbench choose the boot source. By abstracting the concept of the boot source from the tests, it becomes easier to adapt the tests to new environments and to introduce randomness in testing.

> *As an architectural bias, making smaller classes with minimal assumptions can lead to code that is more adaptable.*

Where is adaptation likely to happen?

A fruitful place for adaptation is at the physical interface, or connection layer, of the chip. This is because an external standard (such as PCI or USB) often defines the behavior there. Note that programming the protocols is not as clear cut. This is where the trade-offs between flexibility and complexity, and the use of minimal assumptions, come into play. (Trade-offs are discussed at length in the chapter on OOP connections; we'll just start thinking about this issue here.)

As an example, monitors on a bus might use events to coordinate with higher-level checkers. Because the monitor does not know whether a checker is waiting on an event, the connection is relatively weak and thus the code might be made more adaptable in this case. As a result, this same monitor class could be used in several different environments, each of which is interested in a different set of events.

---

[1] A serial computer bus invented by Philips that is used to attach low-speed peripherals to a motherboard, embedded system, or cell phone. The name is an acronym for Inter-Integrated Circuit and is pronounced I-squared-C.

Another good area for adaptation is when you exercise a feature of the chip. Conceptually, the code that has to be executed to verify a subsystem is independent of where that subsystem is located. In this case, the trick in maximizing adaptability is to minimize the assumptions the code makes about the testbench environment. If a common framework can be used, including what a testbench provides and the application of a few global objects, it is more likely that tests can be adapted.

Initially, concentrate on the lowest levels of abstraction for creating adaptable code. Then, concentrate on the part of the test that exercises a chip protocol or feature.

Separating Interface from Implementation

VHDL encourages a separation of the code interface of a module from its implementation. This separation is a bit more difficult in the Verilog language, which is usually limited to various stub implementations of modules. SystemVerilog is like VHDL in that it supports a separation of "what you can do" (code interface) from "how it is done" (implementation).

Why is this important? By separating these two concepts, one makes a a design not only simpler, but also more adaptable. The design is simpler because the roles of user and implementor are separated. The user of a class is concerned only with the code interface. The implementor, on the other hand, is concerned with how to accomplish the code interface. Neither task is simple, but developers can concentrate independently on what they need to do.

The code is more adaptable because the implementation may change or evolve as the project goes on, or as the code is used in other projects. However, the code that simply uses the class does not have to change. This is particularly important when class inheritance is used, as explained in the next section.

The separation of the code interface from its implementation is an example of the defining of roles and responsibilities of the code. This is a common theme in this handbook, one that is explored in depth in the next chapter.

By separating the code interface from the implementation, we make code that is both less complex and more adaptable to different situations.

Code Interface, Implementation, and Base Classes

Part of "thinking OOP" involves using classes to express common behavior. In fact, expressing one class as a derivative of another is the main mechanism of OOP. But how can this help you write better code? It does so in two ways: one lets you clearly specify a code interface to be implemented, and the other lets you reuse implementations of existing classes. (Don't worry if this issue seems a bit hazy; it is discussed in detail in the chapter on coding OOP.)

To create a class that is to be used in a code interface, just use pure virtual methods. For example, to create a top-level code interface, you might have something like this:

```
virtual class test;
  'PURE virtual task build ();
  'PURE virtual task run ();
endclass
```

Then, specific tests implement the build and run methods, as appropriate for the test at hand. Note the following examples:

```
class dma_test extends test;
  virtual task build ();
  //code here to build the test...
  endtask
  virtual task run ();
  //run the classes built by the above method
  endtask
endclass
```

The `dma_test` can add other methods as it sees fit; but because it derives from `test`, it must implement the `build()` and `run()` methods.

The other benefit of using inheritance is to save time during implementation. The idea is to have a base class implement the bulk of some common code. Then classes are inherited to implement the specific parts that cannot be generalized.

For example, suppose we had a protocol that can be expressed in a common algorithm, but the final wire-driving mechanism were specific to an implementation of the protocol. Here's how the base class might look:

```
virtual class the_protocol;
  extern task do_send(); //implemented in a source file
   'PURE protected virtual drive_wire_(bit logical_value);
endclass
```

Now, a class can inherit from `the_protocol` and provide the actual driving of the wire.

Class inheritance is a main part of OOP. Use it to specify a code interface, as well as to reuse class implementations.

Summary

We have started "thinking OOP." We talked about how verification is complex, and how this complexity must be managed. We talked about the goal of creating flexible, adaptable code, and how achieving this goal may complicate the code.

We got into coding a bit by talking about the separation between a code interface and its implementation. We also touched on the subject of base classes and the different ways they can be used.

For Further Reading

- The sections "A Complex Solution" and "Accidental Complexity vs. Implementation Complexity" are drawn from the landmark paper, "No Silver Bullet: Essence and Accidents of Software Engineering," by Frederick P. Brooks, Jr.

- The concept of team productivity and its variability is well documented in *Peopleware: Productive Projects and Teams,* 2nd edition, by Tom DeMarco and Timothy Lister.

- One of the "lessons learned" is that the social elements of the team can affect the code you write. Although this is currently an important topic of the software domain, we have decided to stay focused on code techniques. However, there are several good books on the subject of social elements if you are interested.

  The grand-daddy classic is *The Mythical Man-Month*, by Frederick Brooks.

  Organizational Patterns of Agile Software Development, by Jim Coplien and Neil Harrison, is a good analysis of several years of software projects.

  Lean Software Development, by Mary and Tom Poppendieck, looks at how proven "agile" manufacturing techniques can be applied to software development.

Designing with OOP

That's right!" shouted Vroomfondel. "We demand rigidly defined areas of doubt and uncertainty!"

The Hitchhiker's Guide to the Galaxy, *by Douglas Adams*

What is design? How do you go about designing with objects? As far as a CPU is concerned, you don't need objects. You might as well put all the code into one big function—but that code would be really, really hard to understand. So we break up a single huge solution into a number of smaller, understandable, and rigidly defined pieces. This is the essence of design, regardless of the language.

When we design with OOP, this "breaking up" is really just the construction of a network of classes. However, before we dive into making classes, we need to understand the OOP design bias. OOP designs tend to focus on roles and responsibilities. These roles and responsibilities get broken down further into smaller networks of classes. Ultimately, we wiggle wires to communicate with the chip.

In this chapter we introduce some basic guidelines for design. We also talk a bit about some common design mistakes and how to avoid them.

Overview

Design is so intertwined with coding that it's artificial to separate the two. Academic textbooks explain that first you architect, then you design, and then you code. In the real world, however, these steps all get jumbled together. We tend to do all three at once, having a general idea of what we want to do, then refining and changing our idea as we start to code.

Designing with OOP is no different. We talk about "paper napkin" or whiteboard designs. We prototype and refine class interface files and talk about what each class should do. Just as important, we talk about how the classes interact and exchange control and data.

Design is messy, but designing with classes can be a little cleaner. This chapter provides some general guidelines that can help with this inherently untidy process.

Keeping the Abstraction Level Consistent

A key evolution in programming came about when we started to talk about "abstraction levels" in a design. This is somewhat expected, because humans are abstraction machines. A child can recognize a "chair," from the folding chairs at school to the hydraulic ones we use at work, and most people can operate a car, regardless of the make or model. Our mind's ability to abstract away the details of an object or process is directly applicable to programming. We can solve a complex design by using abstractions, from the big-picture operations at the top, down to the wire protocols at the chip interface level.

To put this in fancier terms, at any layer in a design there is an associated *scope of concern* and an appropriate *level of detail*. A scope of concern is the role of the task. The level of detail is the responsibility of the task.

At the top level of the verification system, the scope of concern is the entire chip, its configurations, and the traffic that will be applied to the chip in testing or real life. Here, the level of detail should be very small. In other words, the test should consist of "big" objects, such as the test and testbench, and have no minutia. At the other end of the spectrum, at the bus functional model (BFM) level, the scope of concern should be very small (for example, a handful of pins and wires), but the level of detail should be high (for example, the precise sequencing of those pins to implement the protocol). This is shown in the diagram below:

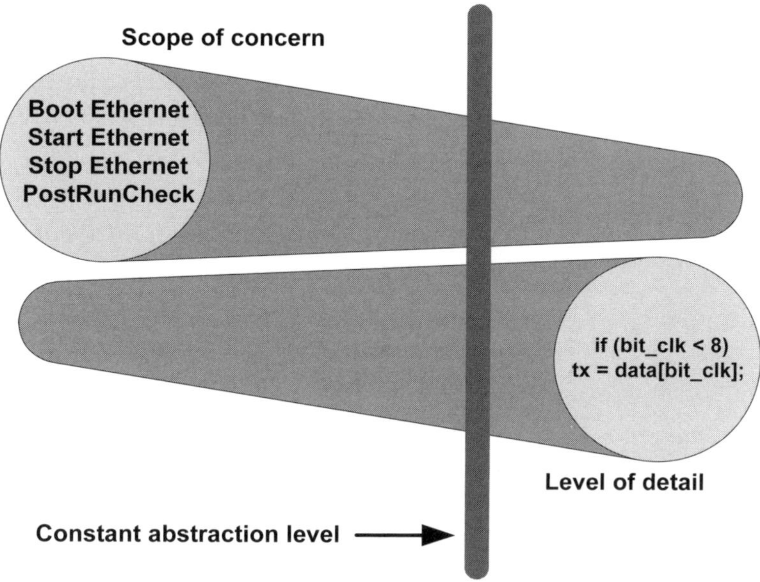

Unfortunately, changes in the abstraction level within an algorithm cause confusion and increase the complexity of the code. For example, shown below (actual code from a coworker) is a top-level algorithm with two shifts in abstraction level. See if you can spot the shifts.

```
task main_process_loop(uint32 num_transfers,
                       IO_BFM io_unit);
  for (uint32 i=0; i < num_transfers; ++i) begin
    BFM_command command = new ("a_command");
    assert (command.randomize());
    io_unit.top.driver.process_command (command);
    for (uint32 j=0; j < 300; ++j) begin
      @ (posedge (iface_.clk));
    end
  end
endtask
```

The first shift occurs in `io_unit.top.driver`. This is because the engineer trying to understand the algorithm must now understand two more classes, `top` and `driver`. It probably would have been clearer either to provide a `process_command()` method in the `io_unit`, or pass in only the driver to this function. Multiple periods in an identifier are usually a cause for concern.

The other shift is in the `@(posedge(iface_.clk))` statement—which is too detailed for the rest of the algorithm. Is it really necessary to worry about clocking at this level?

> *Object-oriented programming is all about using abstractions! Be sure that a class provides some well-defined service at a fixed level of detail. The implementation of the object is one level down in abstraction, and probably uses lower classes to get its job done, and so on ad infinitum.*

Using "Correct by Construction"

As we have mentioned, building a verification system creates a large and complicated network of classes, instances, and conventions. It's not always obvious how to put these building blocks together. However, the SystemVerilog language provides strong type checking that can help to communicate the "intent" of the construction. This strong type checking can give clues as to what classes can go together and how they go together. You should strive for systems that, if they can be put together (compiled), are correct. This technique is called *correct by construction*.

Base classes are often used to show intent. A base class can be used to specify a required code interface, or to manage a list of homogeneous objects (whose actual types are inherited from the base). For example, a base class called `checker` might be used to indicate that a class has checking type behavior and has a concept of when it is done, as follows:

```
extern virtual class checker;
  'PURE virtual task wait_for_completion ();
endclass
class ethernet_checker extends checker;
```

```
   extern virtual task wait_for_completion ();
endclass
class pci_checker extends checker;
   extern virtual task wait_for_completion ();
endclass
```

In this case, both the `ethernet_checker` and `pci_checker` can be assumed to have some action that takes time and has a completed concept. (Note that in this example, it is probably not appropriate to have a list of `ethernet_checker` and `pci_checker` objects. They are unrelated in function, and are related only by inheritance.)

Here is an example that encourages the building of a list of base class objects:

```
typedef class data;
package pci_configuration;
   typedef enum {in, out} request;
endpackage
virtual class pci_endpoint;
   'PURE virtual task handle_data_request(request r,
       data d);
   'PURE virtual task handle_data_completion ();
endclass
class configuration_endpoint extends pci_endpoint ...
class address_endpoint extends pci_endpoint ...
class power_management_endpoint extends pci_endpoint ...
```

PCI can be viewed as having several types of endpoints, all of which respond to `in` or `out` requests. In addition, each endpoint has an associated action after the data have been sent (or received). Thus, it makes sense to have a common base class, and probably a list consisting only of pointers to the base class, `pci_endpoint`. The actual data in the list will be of the inherited classes.

Enumerations can also be used to show intent. They can be given names that match the chip's control/status register (CSR) field that they represent and can have values that directly map to the chip, as in the following example:

```
typedef enum bit[12:0] {window_4K = 0x100,
   window_64K = 0x101, window_1M = 0x102} window_size;
```

Be careful, because enumerations, once defined, cannot be extended. Furthermore, code that uses a `case` statement on enumerations is a possible sign of design trouble, and the use of enumerations may need to be reconsidered. The warning signs are major amounts of code in the `case` label, or multiple `case` statements on the same enumeration. In this situation, an enumeration is being used as a control flow mechanism, not as a simple data mapping. It's possible that the enumeration is better represented as a set of classes. Using an `enum` with a `case` statement is not necessarily wrong, and it certainly is useful as a name for a bit pattern that the chip understands, but be aware of potential problems.

This is just a quick tour of the concept of "correct by construction." (Many examples of these techniques can be found in the Part IV of this handbook.)

Be careful with systems that must be validated at run time. While some parts of a system will need to use run-time checking, this should be the exception, not the rule. Systems that use run-time checking make it much harder for others to understand your intent.

Base classes and enumerations are good mechanisms for making it easier to see and enforce how a system can be put together.

The Value of Packages

When a large amount of code is developed, there will be enumerations or constants that have the same name. If you had access to the source code, you could standardize the colliding names. However, changing the code is more difficult when the code is from another work group, division, or company. SystemVerilog helps minimize this problem by providing a feature called a *package*. A package is like a class, in that the methods (and data) in the package must be accessed with the name of the package. By using packages, you minimize the number of global names, because the previously defined global functions are all in packages, and now only the package identifiers themselves are global. This decreases the probability of a collision.

Packages are useful for grouping related parameters. A good design bias is to place the related parameters for each verification component in your system in their own package. For example, in a file called `lcd_parameters.svh`, you might have something like this:

```
package lcd_parameters;
   typedef enum {TN, S_IPS, MVA, PVA} type;
   typedef enum {r_60, r_70, r_80, r_100} refresh_rate;
   typedef teal::uint32 pixel_rate;
   const¹ pixel_rate max_pixel_rate = 'h5551212;
endpackage
```

You can use the enumerations and constants in a package by specifically naming it (for example, `lcd_parameters::<id>`) or making it implicit by means of the `import` keyword. Below is an example.

```
'include "lcd_display.svh"
import lcd_display::*;
//now lcd_display enumerations may be used
//without qualification
task lcd_function();
   lcd_display::refresh_rate r = r_60; //can be specific
   refresh_rate r1 = r_60;             //or not
   pixel_rate pr = max_pixel_rate;
endtask
```

Be extremely cautious of putting an `import` clause in a header file. It is almost always a mistake, because every file that includes your header file will inherit the `import` clause's scope. In addition, every file that includes the header file that in turn includes your header file will now have the `import` clause—and so on, with possibly unintended consequences. The authors have first-hand experience in trying to undo this technique. The task was not pretty.

Another use for a package is to wrap a related set of global objects and functions. The interface is just a collection of functions, but wrapping them into a package creates what is called a *singleton*. While you want to minimize the number of singletons in a system, they are necessary and correct for those areas that represent global resources. Singletons are discussed more in the Coding OOP chapter.

[1.] Within a package in SystemVerilog, a `const` is similar to a parameter.

Packages are very useful for grouping related enumerations and constants. Be careful with the `import` *clause in header files.*

Data Duplication—A Necessary Evil

There is an inherent trade-off between minimizing the complexity of connections among verification components and duplicating the data passed among them. The looser the connection, the more likely that there is a duplication of data. This is talked about in detail in the next chapter, but we'll discuss it as a design bias here.

Let's look at an example of duplicated data. Consider a DMA[1] chip's view of data versus a checker's view. A processor sends to the DMA engine the source address, the destination address, and the length of the transfer. When the DMA completes, selected memory contents are stored in an object that contains a start address, an end address, a length, and a completion status for the transaction. The chip synchronization mechanism for the monitor is probably an interrupt. The class and the chip memory both contain the data, so we have a data duplication situation. In this design the data duplication is good, because checkers can use the more abstract fields of a class, instead of reading memory directly. Also, the code is more portable, because the concepts of source, destination, and length are abstracted away from the actual memory layout.

In the previous example, the data representation changed from chip memory to fields in a class. However, there is another place where data are duplicated. This is a more subtle, yet useful, design technique.

We think nothing of calling a function and passing it an integer. We know that the integer's value will be copied. But what if this integer were the result of some computations in the function making the call? Now the situation is not so clear. Why? Because this integer is not just a value. It represents the result of an algorithm that computed some data. The fancy term for what this integer represents is *derived data*. This is a normal situation, and it is used all the time when abstraction layers are crossed.

---

[1] Direct memory access.

In fact, derived data are common in multilayer protocols. Many protocols have a physical layer, which deals in bits and words. Then they have a transport layer, which deals in packets. Some protocols even have a third layer, which deals in higher-layer transfers. As each layer hands off to the next, data are copied. The concept of derived data also exists when abstraction layers are used in verification.

For example, consider the physical interface for Ethernet Media Access Control) MAC. Suppose a verification component called MAC supports both Media Independent Interface (MII) and Reduced Media Independent Interface (RMII), and expresses this in an enumeration. These two interfaces are quite close, with RMII being a reduced set of the MII interface. In order to simplify the control connections between the MAC and the physical layer, the connection could be a single bit set to one for "full," to distinguish it from "reduced," where the bit would be zero. Although the information in a single bit is weaker and simpler than the passing of an enumeration, the bit is more appropriate to the lower-level abstraction.

Be aware of the same data being in two places at once. This generally happens across abstraction levels. Be very aware when derived data must be refreshed.

Designing Well, Optimizing Only When Necessary

When hardware is designed and implemented, efficiency is critical to its ability to be built (meeting timing, power, and size constraints). However, in software the emphasis is often on minimizing mental complexity. This is natural, as increasingly complex control and data structures can be built relatively quickly with software.

In verification you must focus on clarity in your overall design and implementation.[1] This does not mean, however, that you can ignore efficiency.

---

[1] The authors assume, of course, that the verification of the chip is the primary concern.

The speed of a functional simulation almost always depends on just a small percentage of the actual code. You should optimize the code for simulation speed only after you have finished profiling it thoroughly. Premature code optimization leads to confusion. While it might run a simulation a little faster, it will more often than not slow down the project by being harder to understand and use.

> *Get the design working and understandable first. Then figure out what needs to be optimized—and possibly made more complicated.*

Using the Protocol, Only the Protocol

The verification test system must both apply stimuli to the chip and then check to make sure the chip generated the correct response. Consequently, the verification system is aware of *both ends* of a data path—an awareness the software does not have. In a simulation, creating both ends of a data path is necessary, but it can lead to sloppy code that may mask chip problems. Be aware of what information the protocol provides, and use only that information for checking.

For example, the authors helped test a USB device. The original verification tests preloaded the chip with the "correct" data. Then, when the data were needed, as specified by the protocol, they were sent by the chip to the checker. The checker confirmed the data and all was well. Upon closer inspection by the software team, a critical parameter was found to be missing: the length of the response. By design, this parameter was being dropped by the chip, but because the verification code already knew the "right" answer and length, the design error was missed. Had the verification code been cleanly separated into a *generator* (the requestor) and a *checker* (the interpreter) of the request, the error would have been caught.

Another example concerned a DMA checker. The DMA checker required the verification components to register all the memory accesses that the chip would make. At the end of the test, the DMA checker would make sure there were no unmatched writes. What the checker failed to test, however, was the result of overlapped memory writes. There was a bug in the cache coherency unit that was masked because the tests did not write to the same address twice.

Minimize the assumptions made across different parts of the verification system. Strive to use only the "knowledge" that the production software has through the feature or interface.

Verification Close to the Programming Model

Verification is both similar to and different from the production software that will run on the system. However, the closer the verification architecture is to the software, the better the chance you have of finding errors in the programming model, with respect to condition status registers (CSRs), hardware algorithms, and so on.

So while it is true that the software you use for verification is inherently a lot more detailed than the production software, you should still strive to keep your test algorithms as close to the production ones as possible. This design bias not only helps the software and hardware folks communicate better, but it can also help shake out register/memory/interrupt design and programming model.

For example, the authors had a project where the design of the DMA part of a chip was changed after the verification team had their code reviewed by the software team. Yes, this change was humbling. Specifically, the chip used a bit to restart a DMA channel. The original design had made the transition of zero to one be the enabling action. This meant that the software would first have to set this bit to zero (it was a one from a previous enable!), and then set it back to one. Once the software team was aware of this issue, they asked how hard it would be just to restart with a write of one, even if the previous value were a one. This was done, making the DMA enable more intuitive to the software team. Another change concerned the ability to write the DMA offset and index. The original design allowed only a reset of the index to zero, which made resending a DMA buffer clumsy. The software had to recopy all the buffer descriptors in memory to the zero offset address. The hardware was changed to allow the index to be written whenever the DMA unit was paused.

A final change was in the address map. Previously, the address space for the DMA was contiguous, which made sense to the hardware team. However, because the various I/O subsystems were features that end users paid for individually, the software was separated according to I/O subsystems. Because of this division, there were two different address spaces to manage: one for the I/O subsystem, and another for the DMA. We changed this design and moved the specific DMA channel's CSR address space to the I/O subsystem that it served. While this change seems (and probably is) simple, it shows that the hardware and software teams viewed the chip differently. Working more closely together improves the overall quality and helps reduce total project time. Furthermore, using a software bias allowed the two teams to look at each other's code without too much complaining.

Try to design your verification system to use the same algorithms and subarchitecture as the production software. That way, you can catch clumsy or conflicting programming models.

The Three Parts of Checking

The major thrust of verification is checking the operation of the chip. We send data in, or turn on features, and check that the chip produces the correct response. A fancy term for the transmission of data and the enabling of features is called *stimulus*. As a design bias, it's a good idea to separate the generation of the stimulus from the acting on and checking of the stimulus. We'll look at this further in subsequent chapters, but for now let's look at the *checker*.

Once data have been injected into the chip, the checker recovers the actual data output from the chip's I/O and confirms that everything is as expected. There are three parts to this process and, to promote adaptability, the implementations of each part should be kept separate. The success of code adaptation can often be traced to how well these three parts, summarized below, can be shaped to fit a new environment.

- The first part is the gathering of the data, usually by means of a monitor. A monitor triggers on changes in some I/O, interrupt, or FIFO level and converts these wire changes into verification objects. By having a monitor that is separate from the checker, you can change how the data are gathered without affecting the checker. Also, by converting the data from wire changes to integers and classes, you elevate the data by an additional level of abstraction.

- The second part is a comparison of the actual data with the generated data. This can be accomplished by providing the method `equal()`. (Part IV of this handbook shows examples.)

- The third part of checking uses the result of comparison and provides an indication of expected or erroneous behavior. The simplest example is an error message or "check passed" message, followed by a printout of the data.

In the checker class, be aware that the form of the data generated may not be the same as the form received by the monitor. This is because data packets may have been combined or split for a variety of reasons (for example, because of the protocol, error correction, or some other transformation). Creating an appropriate level of abstraction for a checker can be difficult.

Sometimes there is more than one level of checker. This is common in multilayer protocols, such as PCI Express, Ethernet, ATM, and USB. Again, keeping the levels separate improves the code.

Sometimes there may be several monitors on the same chip I/O. This is common because in the verification code a one-to-one relationship between monitor and checker is the simplest, while in the HDL there are no simulation limits on the number of monitors on a wire.

The checker should also check for dropped or missing data. The easiest way to do this is at the end of the test, but it's probably better to use the latency of the chip as a filter, and report errors as soon as possible.

Checking consists of three parts: gathering data, making a comparison, and acting on the comparison. Separating these three parts is a "flexibility vs. complexity" trade-off.

Separating the Test from the Testbench

It is common to have a few top-level parts to a verification system. These include the HDL top, the testbench, and the test. The authors advocate for another top-level component, the verification top (see the Layered Verification Approach chapter earlier on). This section addresses the roles and responsibilities of each part.

A major theme of this chapter is to show how clear roles and responsibilities create simpler code. In general, for a single chip, try to have a single verification top, one or a few HDL tops and testbenches, and many tests. Because the verification top is the top-most component of any test, always calling the same "dance," it consistently executes the same steps of creating, configuring, running, and shutting down the test.

The testbench is responsible for setting up the transactors, monitors, and generators (under direction by the test), and building the global services. The HDL top is responsible for the HDL wrappers around the chip, and includes clocks, reset logic, and pin wires. In addition, the HDL top probably contains muxes, assigns, or Verilog `tranif`[1] statements for connecting verification interfaces to the chip. It may also have interfaces for "power on" (to communicate with the boot I/O devices and enable the initial state configuration of the chip). The test is responsible for specifying the required verification components, traffic patterns, and verification configurations, as well as what the "run" part of the test should do.

Having a single testbench for a specific chip or system is useful, because it sets up a common environment for all tests to use. This increases the adaptability of the individual tests and verification components. Also, because unit testing differs from full system testing, it may be necessary to have multiple HDL tops; however, effort must be taken to ensure that as many of the tests as possible can be run with any of the HDL tops.

Because a test must specify what configuration it requires, there is some communication between the test and testbench, possibly before the chip can be brought out of reset. This means that, while the verification top creates the test, the test may make several calls to the global objects,

---

[1.] The Verilog primitive `tranif` connects two bidirectional wires.

possibly including a verification components manager, to configure the testbench.

As an implementation detail, the test may be created by a *factory function*. (Factory functions are explained further in the Coding OOP chapter.) This function, usually implemented in the source file that contains the test, returns a base-class test pointer. The reason for using a factory function is to incorporate unit-level tests into a full chip test, or many tests into a single meta-test.

Sometimes this base test object contains all the verification components and a basic structure for the test. This can be useful, but be aware of all the types of tests, and don't make the base test too complex.

A test can have a few standard components: the dance, the testbench, the HDL top, and the test. Only the implementation of the test should change.

Summary

We have started down the path of using OOP in a verification system. We talked about the main theme, creating roles and responsibilities by using abstraction. We talked about the common design biases used when we design a verification system.

You probably are still surrounded by clouds of uncertainty. This is understandable. The next chapters are more specific, talking about making classes and the different ways to connect them.

For now, however, know that designing with OOP is about defining roles and responsibilities and making levels of abstraction, a "layering" for which there are many examples in our everyday lives. To achieve your own design objectives in silicon, use your experience to guide the process.

For Further Reading

- To help you think about how to construct a system with abstraction levels that are logically consistent, a great book is *The Design of Everyday Things,* by Donald A. Norman. Though it does not deal with code or high-tech, it is great for thinking about how someone else might develop a mental model of using your code.

- On the topic of connections between levels of abstraction and within a level of abstraction, *Software Engineering: A Practitioner's Approach,* by Roger S. Pressman, has several pertinent sections. (The fancy term for these connections is "cohesion and coupling.")

- Bjarne Stroustrup also provides a concise discussion of abstraction in *The C++ Programming Language*, section 24.3.1: *"What do classes represent?"*

- The concept of "correct by construction" is from Edsger Dijkstra, a pioneer in formal languages, specifications, and proofs for computing. This concept is often used in formal verification, but it is adapted here to show how design intent can be communicated.

- Regarding premature optimization, the original quote is from Tony Hoare: *"We should forget about small efficiencies, say about 97% of the time: premature optimization is the root of all evil."* This was quoted in Donald E. Knuth, *Literate Programming* (Stanford, California: Center for the Study of Language and Information, 1992), 276. A web search can provide further references.

OOP Classes

Experience is a dear teacher, but fools will learn at no other.

Benjamin Franklin

Coming up with the appropriate classes for your project is an experience-based effort. In other words, the authors made many mistakes in the beginning. To help you in designing classes, we have collected experience from our previous efforts.

So do what the authors did when they learned SystemVerilog: find examples, copy, and paste!

This chapter introduces the thought process for creating classes, to answer questions such as these:

- How do I determine what is a class, and what is a method?
- How should I handle global functionality?
- What can inheritance do for me?

Overview

Classes are fundamental to writing in an object-oriented language. But how do we decide what is a class? We have talked about thinking in terms of layers of abstractions. We have talked about roles and responsibilities. The next thing is to start to name the classes and their responsibilities. This is not as hard as it sounds. For one thing, you make classes as you feel they should be, and there is no right or wrong way. Let each class do what feels right to you. There will, of course, be some spirited team discussions.

Once you decide on some classes, you can "wire up" instances of classes pretty much like you create and "wire up" modules in hardware. Unlike hardware, however, classes can have more "electricity." When designing hardware, you are restricted to connecting blocks through wires or signals, but with classes you have the ability, among other things, to have pointers to other class instances or call virtual methods. This is the topic of several sections.

This additional freedom is where the electricity comes in. This is good, because it helps you solve complicated verification problems. As with any technique, the challenge is to use the appropriate amount of electricity.

Not that everything has to be a class. As we learned in the previous chapter, SystemVerilog supports tasks and functions in packages, and for many situations, this is appropriate. The section in this chapter on Global Services talks about various ways to use global functions.

Defining Classes

As object-oriented programming has been around for many years, there have been many different attempts to explain how to define good classes. In the end, it comes down to the usual way one learns: copy examples, change the example a bit, and eventually start writing your own code. After some time you will find your own way to "ride the OOP bicycle." That's the reason this handbook presents lots of code snippets and examples.

That said, a common way to define classes basically just follows the old grammar school rules for writing a good sentence: make a class for each noun in your design, and make a method for each verb. This means that each block in your whiteboard design becomes one or more classes. Drawing the lines between the blocks is a bit more tricky. At some level these lines represent method calls, but they can also be classes themselves. That's the great thing about OOP compared with HDL design; you can use a variety of alternatives (more language constructs, more techniques allowed by basic constructs) as you discover problems in implementing the initial "obvious" whiteboard class design.

It is promising that the industry has finally settled on (more or less) standard names for the most common classes. Names such as *generator*, *BFM*, *monitor*, *driver*, and *checker* have become somewhat standard in their meaning.

> *Making classes becomes only easier with experience. First clone and modify existing code.*

How Much Electricity?

SystemVerilog is a language that allows a reasonable amount of "electricity" in the code. The electricity in a piece of code is a measure of how complex the code is. Recall that complexity is inherent in our world; it's the management of complexity that prevents complex code from

becoming complicated code (remember that complex is okay, complicated is bad).

The goal of any OOP system is to design classes that have minimal
electricity. Anything more is just unnecessarily complicated.

At the lowest level of electricity are *defines* and *macros*. In order to figure
out the code, you have to figure out the value of the define. The next
level is "if" tests, which are dealt with in the chapter on Coding OOP.
The Verilog language has these capabilities as well.

Classes

The next addition of electricity consists of *classes*. The Verilog language
has the *module* concept, which is pretty close. Both modules and classes
unite data and algorithms. SystemVerilog and Verilog differ in that in
the former, the data can be classes, whereas in the latter, the data can
only be wires and registers. Furthermore, in Verilog modules represent
silicon, whereas in SystemVerilog, classes represent a wide range of
concepts.

Packages

Related to the concept of classes is *packages*. This is somewhat similar
to the *package* concepts of VHDL. Packages are useful in that they group
related enumerations, global functions, and constants together loosely.

(This was discussed in more depth in previous chapters.)

Pointers and virtual functions

Another increase in electricity relative to operator overloading is found
in *pointers* and *virtual functions*. These relatively simple features have
profound implications. The number of techniques that can be realized
with this electricity has spawned numerous books and papers. Remember
that at the implementation level, virtual functions are pretty much just
plain old pointers to a task or function. You may have used this technique
before in other languages.

*SystemVerilog provides many levels of electricity. Use
minimal electricity in your designs.*

Global Services

OOP-based design is about using abstractions and defining roles and responsibilities for specific classes. By using layers, as described in the Layered Approach chapter in Part I, you can simplify the design and set up a network of classes. For example, a monitor and a checker have a *neighbor* relationship; the monitor takes data from the chip, and the checker checks the data.

However, some roles and responsibilities are related to a large number of other classes. Activities such as memory reads and writes, control and status register (CSR) reads and writes, interrupt vector handling, and message logging can reasonably be expected to be available to all classes. There are many more examples. Roles and responsibilities that are available to all classes are called *global services*.

A logical way to create global services would be to use a class for each service. Then, pointers to these global service classes could be passed to all other classes. However, in practice this is often a clumsy approach. Passing the global service objects to the majority of all classes clouds the code and adds little to the real information of a design. Used with restraint, though, global service objects can extract the common components from the mental baggage of learning a new design structure.

Package it up!

As we have seen, the authors prefer to use a *package,* or sometimes static methods of a class,[1] to express the intent implicit in a global service. As a result, any class can include the header file and then use the service.

For example, consider a memory package such as the following:

---

[1] Note that not all simulators support static functions yet.

```
//in the file memory.svh ...
package memory;
   extern task write (uint32 address, uint32 data);
   extern function uint32 read (uint32 address);
endpackage
```

Note that `read()` and `write()` are memory functions. For this service, any class can include `memory.svh` and start accessing memory through these functions; no special rules or objects are needed. Each class accessing the memory functions still needs to use the `memory` package, which gives the reader a way of finding the source of these functions.

Therefore, using packages can simplify a verification system while keeping the source available in a central place. In addition, any class can access memory without having to know about how the memory is implemented.

Static methods

Another way of presenting a global service is to use *static methods*. Using the same memory example as above, you could instead declare the global services in a class like this:

```
class memory;
   extern static task write (uint32 address, uint32 data);
   extern static function uint32 read (uint32 address);
endclass
```

The way to access the global service in this example is pretty much identical to using a package, and the end results are similar.

Singletons—A Special Case of Static Methods

Instead of using all static methods in a class, you can use a single method to get the one instance. After that single method is used to get a pointer to the single instance, the accesses show up as with any other object. This can be useful if you want to communicate that there is a class instance performing the work. On the implementation side, it can make the different implementations of the (previously declared) static methods easier. At any rate, using this technique implies that there is more electricity than is the case with just a package or static methods. Let's look at the memory example again.

```
class memory;
  extern static memory get ();
  'PURE virtual task write (uint32 address, uint32 data);
  'PURE virtual function uint32 read (unit32 address);
endclass
```

The single static method, get (), is used to get the single instance of the class. This technique is called a *singleton*. Notice that the read and write methods are virtual. Let's look at how this technique might be useful in a burst memory class.

```
typedef uint32 data_list[];
class burst_memory;
  extern static burst_memory get ();
  extern virtual task burst_write (uint32 first_address,
                                   data_list data);
  extern virtual function data_list read (uint32
     first_address);
  extern virtual task write (uint32 address, uint32 data);
  extern virtual function uint32 read (uint32 address);
endclass
```

The implementations of the read and write methods are probably simpler now that a singleton is used. Otherwise, the read and write methods would both have to deal with gaining access to the correct memory.

Packages or static methods?

So why prefer one technique—packages or static methods—over the other? The reason the authors prefer using packages is that packages are more reasonable than static methods, because they are less complex.

With static methods in a class, you run into trouble when you need to add a new static method. Doing this requires that you inherit from the original class and add the new static method. Although using static methods for global services is useful for more component-like services, using a package is often easier than using a static class method. Nevertheless, despite this advantage of simplicity, there are times when you want to bring to the code interface the fact that the implementation is an object. That is when singletons may be appropriate.

Using packages is a good way to provide access to global services, and it is less complicated than using static methods. Although singletons are a good way to implement a global service in a code interface, be aware of where the object is created.

Other considerations

One last point before we move on. A global service, at the code interface level, implies a logical single service. The actual implementation of this service, however, may be vastly different behind the scene.

For example, to write to different memory addresses, several different objects may be needed. This is because the memory space is probably spread out across different chips—or at least across different interfaces of a chip. Also, if you fold the register access into the memory access, different register banks may communicate with different chips. Finally, you may want front- and back-door access for the memory and registers, which would require different objects to do the implementations. This can all be an implementation detail for the end user. (To see an example of this, refer to the implementation of the `memory_bank` class in the `teal_memory.svh` file of Teal.[1])

Another technique is to implement a singleton as a list of filters. This is useful for purposes such as filtering logging messages. For example, a logger might look to the classes that use it as a single object. The implementation, however, may want to use a linked list of objects, where each object is given the chance to modify the log message. This is a powerful technique. (To see an example of this, refer to the implementation of the `vlog` class in the `teal_vout.svh` file of Teal.)

Often there are global services in a verification system. Use either packages or static methods to express them.

---

[1] Available at www.trusster.com.

Class Instance Identifiers

When you start printing log messages, you have to decide how to identify the object that is printing. This object identification provides a way to trace the object through the system.

There are at least two techniques for identifying an object. One uses a string (often the name of the instance), and a second uses some sort of sequence counter. In practice, both techniques, as well as their combinations, are used.

Strings as identifiers

As you move up in abstraction level, it will at some point be better to have names for objects. These are most often placed in the constructor of a class, as follows:

```
class fabric;
  extern function new (string name);
  local string name_;
endclass
```

With the class using a string for a name, it should be easier to print useful status messages. Because the name is passed in, it is easier to make a unique name reflect its use in the chip.

Static integers as identifiers

Sometimes it's too cumbersome to use a string as an identifier for an object. Also, the name, while unique, does not indicate a sense of sequence between consecutive objects.

For example, often a generator creates a sequence of objects, as in, say, a number of Ethernet packets. In this case, it may be appropriate to name the instances with an incrementing integer, such as `packet_1`, `packet_2`, and so on.

Having sequence numbers can be useful as a triggering mechanism for trace types of logs, or for postprocessing the log file. This is done by making the counting integer `static`; this declares a class-level "shared" integer, and increments it for each instance of a class.

Here is an example:

```
class data_packet;
   function new (); my_id_ = ++id_; endfunction
   local uint32 my_id_;
   //count of data_packets created,
   //starting with 0
   static local uint32 id_;
endclass
```

Then, in the `data_packet.sv` file, you can use the `id` in the name of the object.

Combination identifiers

In practice a combination of these techniques is often used. For example, you may want to prefix the sequence number with a name, which can identify the higher-level sequence (such as `short_packet_43`, or `Device_7:_packet_10_Enumeration_Phase`).

> *Identifying an instance should not be an afterthought. Often, a string is sufficient. For sequential instances, a static counter can be useful in tracing an instance in the log file.*

Class Inheritance for Reuse

When you start to name and build classes, there is a tendency to find commonalities in roles and responsibilities. While this is certainly a good thing, resist the urge to define base classes right away. There are many ways to express common roles and responsibilities. Use base classes only when you have experience from several designs, or when you are actually coding and can use base classes to solve a problem. With these caveats in mind, let's look at inheritance and how it can help to make code more adaptable.

In verification we have to drive and monitor the signals of the chip to exercise a protocol with what is commonly called a bus functional model, or BFM. One side of the BFM is connected to the data generators and monitors; the other side is responsible for driving and monitoring the wires of a chip's interface.

It is good practice to separate the actual driver or monitor into two separate code interfaces. One is the data generator or monitor code interface, and the other is the BFM code interface. This is an inherent separation point, because there is a conversion between abstraction layers from "send packet #3" to the actual wiggling of the wires. This separation minimizes the assumptions about how the BFM does its job.

This separation is a good use of inheritance.[1] The authors call the base class that wiggles the wires the *BFM,* and the inherited class that deals with interfacing with the generators and monitors the *BFM agent.*

A BFM base-class example

Consider a chip interface for sending and receiving packets. The base class might look something like this:

```
typedef class packet;
virtual class bfm;
  extern function new (/*interface to the wires*/);
  extern virtual task start (); //start the receive thread
  //This is the driver part of the bfm ...
  extern task send_packet (packet p);
  //This is the monitor part of the bfm ...
  'PURE virtual task packet_received(packet p);
endclass
```

The `packet_received()` is declared as a pure virtual method, so it must be implemented by the inheriting class. This is because the BFM should not know what to do with a completed packet, but just focus on how to drive the data.

---

[1] There are other techniques for this natural separation. You also might just want to own the BFM.

A BFM agent class

Now an inherited class can add channels (one for "received" and one for "send") to convert from packets to commands, and vice versa. The inherited class would look something like this:

```
class bfm_agent extends bfm;
  extern function new (<two channels, virtual interface>);
  extern virtual task start ();
  //start BFM and then thread
  //to get data from the send queue
  extern virtual task process_send_command ();
  extern virtual task packet_received (packet p);
endclass
```

Reusing the BFM class

We could have just lumped the `bfm_agent` and `bfm` into one class. However, with two classes the responsibility of each is better defined, although things can become a bit more complex.

When another project comes along with a chip that has the same BFM interface, but that accepts only a small fixed set of packets, the new test team can still use the `bfm` class and write a simpler `bfm_agent`. Also, because the new `bfm_agent` class is so simple, they might decide to include the checker directly as part of the agent class, and not use a channel at all.

If the team so decides, their project can still reuse all of the existing `bfm` class code by just inheriting from the BFM with a new, simpler, combined generator/checker class. This is why dividing models into layers of classes is a good idea.

Using inheritance to adapt code can preserve working code and still add features.

Class Inheritance for Code Interfaces

The previous idiom of using inheritance for reuse is common. We can take that to the extreme, and instead of reusing the implementation, reuse the code interface only. This is more useful than it sounds. In fact, this technique is very powerful. You can use the code interface of a class to communicate exactly what the classes in the hierarchy can and cannot do. This helps a reader of the code build up a "mental model" of the system. You can only call the methods defined in the base class—no more, no less.

For example, as shown in the Part II of this handbook, a common design for a testbench is to have a top-level procedure that builds a number of high-level, independent verification components (usually related to the major interfaces or features of the chip).

In order to manage the complexity of the verification system, it is important to have common base classes for these major components, so that all the components in a verification system behave in a similar, predictable way. If each major component is built in a unique way, it soon becomes too difficult to manage the overall environment.

Inheritance for a verification component

Let's look at the following `verification_component` class:

```
virtual class verification_component;
  'PURE virtual task initialize ();
  'PURE virtual task start (); //forks threads
endclass
```

Now lots of common verification components can express themselves as an inherited class, like so:

```
class test extends verification_component ...
class bfm extends verification_component ...
class monitor extends verification_component ...
```

By having all verification components inherit from the `verification_component` class, one can understand that each class can at least be initialized, started, and stopped, because these are pure virtual methods that exist in the base class.

Inheritance for a payload code interface

Because using inheritance for code interfaces is so common, let's look at another example. In verification there is often a payload of data that travels through the system. The data must be random, printed out at various times, and compared with the initial data sent in. If the chip had an Ethernet protocol, for example, there could be an inherited class that extends the base payload data class into an Ethernet payload class.

Here is what the base payload class might look like:

```
virtual class payload_base;
  'PURE virtual task report (string prefix);
  'PURE virtual task do_randomize ();
endclass
```

Note that the `payload_base` class has no data; because this is a generic base class with no data, there is no constructor.

The base class declares two virtual methods, one intended for printing and another intended for randomizing the payload of any concrete class. For an Ethernet packet the base class could be used like this:

```
class ethernet_data extends payload_base;
virtual task report (string prefix);
      for (int i=0; i < data.size (); ++i) begin
          teal::vout log = new ("ethernet");
          log.info ($psprintf ("%s data[%0d] %0d", prefix,
            data[i]));
      end
endtask
virtual task do_randomize ();
  'for_each (data_, 'RAND_8);
endtask
extern task put_to_DUT ();
local bit[7:0] data_[];
endclass
```

Shown above is an `ethernet_data` class that inherited from the `payload_base` class. As required, this new class provides specific implementations of the `print()` and `do_randomize()` methods. It also adds a new method, the `put_to_DUT()` method.

The `ethernet_data` class also adds the `put_to_DUT()` method, which is intended to transfer the payload to the chip.

These three methods—`print()`, `do_randomize()`, and `put_to_DUT()`—are the *only* things you can do with an instance of an `ethernet_data` class.

Note that you cannot restrict the use of a base class by inheritance. By definition, anywhere a base class is used, an inherited class can also be used.

> *Inheritance can be used to communicate exactly what an code interface can or must implement.*

Summary

This chapter covered a rather broad range of topics. We talked about how to look at the verification environment and see that the nouns are classes and the verbs are methods.

How much "electricity" your design needs was covered next. The basic levels of electricity can be seen as defines, macros, the "`if`" test, classes, packages, operator overloading, pointers, and virtual functions. Each step increases the complexity, or electricity, of the code. The idea is to use only the minimal amount of electricity needed.

We then moved on to meta-class-level concepts, such as global services and static methods. Here the idea was that sometimes things really are global.

Class inheritance was looked at in detail. Inheritance is an extremely important OOP concept. By using class inheritance, you can both enforce intent and extend an implementation.

For Further Reading

As we stated in the For Further Reading section of the Why SystemVerilog? chapter, there is a growing number of books devoted to coding in SystemVerilog. Skim many, buy a few, and copy code where you can. Web searches can also be useful.

OOP Connections

Oh what a tangled web we weave,
When first we practice to deceive.
Sir Walter Scott, Marmion. *Canto vi, Stanza 17*

Connecting classes together is more important than the classes themselves. How can this be? It is so because, by definition, the connecting of classes involves jumping around in the code base. Managing this is mentally more difficult than simply managing the code within a class. For example, when you see a class name in a method call, you have to think about why the method needs that class. The answer depends upon whether the system is a tangled web or a well-architected series of connections.

Often you have to find the header file for the class and go look at the implementation of the method. In the worst case, the code makes no sense whatsoever, even after you stare at the implementation. In the best case the connections are obvious, such as when a test is passed the class name of a testbench.

Overview

In hardware design we connect modules together and worry about clock domain crossings. With verification, we connect instances of classes and methods together and worry about crossing the threads of execution. In addition, the connections in verification may be temporary (for example, they are used only within a method), or they may be permanent (for example, when a constructor takes in the name of a logger instance and stores it in a data member).

Connections in your code can either form a spider web of complicated and confusing relationships, or they can be a highway, with well-defined points that connect to other roads. Recall that one person's web is another person's highway, so picking the right connection technique may not be universally appreciated. There are many connection techniques and consequently, as this chapter shows, trade-offs to be made.

We first discuss the various types of connections, then look at implementations of these connection types. We then present the simpler connection types first, increasing the complexity of the connections as the chapter progresses.

We first look at how to classify connections. The type of connection is evaluated according to how much information one class has about another. At one end of the information scale, classes have no mechanism to determine whether any other class instances are connected. At the other end of the scale, a class has a pure virtual method that must be implemented to make the connection.

The idea is to build the appropriate type of connection for the problem at hand. Too loose a connection makes code unnecessarily complicated. This because loose connections make few assumptions about the other side, which in-turn makes tracking events harder. Connections that are too tight, on the other hand, may make code harder to adapt.

While these connection techniques are general, some of them can be used between verification components operating in different threads. Why bring threading into the discussion? In verification systems many events need to happen in parallel. This is normally done through threading. With threading, however, comes a set of problems related to accessing common data. How can a thread be given sole access to a common resource? How can one thread synchronize with another? To solve these problems, *thread-safe connections* where created. This chapter will show techniques to cross thread boundaries.

How Tight a Connection?

Once a verification system has been divided into classes, the next step is to think about how tight the connections among those classes should be. As we have seen, this is a sliding scale, with trade-offs in complexity. A loose connection creates good flexibility but more complex code. A tight connection is easier to understand but harder to adapt when changes occur. As a result, each side of the connection must make assumptions about the other.

Let's look at two points on this scale. Consider a generic data generator and data checker. An obvious way to connect these two components is to have one component have a pointer to the other. For example, you could code the connection like this:

```
class data_checker;
  extern task note_data_generated (uint32 some_data);
endclass
class data_generator;
  extern function new (data_checker checker);
endclass
```

Then you can use the checker and generator like this:

```
data_checker checker = new ();
data_generator gen = new (checker);
```

The pointer example above is considered a *tight* connection. This is because the actual name of one class is given to the other class. Tight connections are obvious and direct, yet they are not always appropriate. They make the code brittle and difficult to modify if the assumptions about either class change. Tight connections are, however, the most commonly used and most appropriate for the common interconnections.

Note that the situation could have been reversed, with a pointer to the `data_generator` given to the `data_checker`. However, in practice the assumptions regarding the number of interconnections are not the same. Often there are several different types of generators, yet usually there is only one checker for a given interface or feature. The number of connected instances is something to think about when you connect classes. It is easier to have many objects point to one than the other way around.

The looser a connection is, the fewer the assumptions that can be made about it. For example, to continue with the above example, one could instead use an intermediary object to manage the connection. The authors call this a *channel*. In this case, you can give the channel object to both the checker and generator, as follows:

```
class channel;
  extern task put_data (uint32 data);
  extern function uint32 get_data ();
endclass
class data_checker;
  extern function new (channel expected);
endclass
class data_generator;
  extern function new (channel output);
endclass
```

Then you can use the checker and generator like this:

```
channel a_channel = new ();
data_checker checker = new (a_channel);
data_generator gen = new (a_channel);
```

This is a *loose* connection, because the generator does not know that a checker exists. The generator simply generates data for the channel.

Here is an interesting implementation complexity brought about by our new channel connection. What does the checker do if the `channel::get_data()` has no data? Questions such as this are not necessarily a bad thing; if they are asked in the early coding phase, the resulting code tends to be well thought out. Channels are an important interconnect technique, and are discussed in detail in a later section.

Sometimes tight connections are appropriate, but at other times looser connections give the appropriate flexibility.

Types of Connections

Now that we have talked about the tightness and looseness of a connection, let's look at the two basic types of connections. One is the peer-to-peer connection, the other the master-slave connection.

Peer-to-peer connections

The peer-to-peer connection occurs when a group of modules are all able to communicate with each other at any time. They may be arranged in various topologies, such as a ring, star, or bus. This type of interconnection usually follows a message-passing scheme and can be tricky to debug.

The Controller Area Network (CAN) protocol is an example of a peer-to-peer interconnect. Any device can initiate a transfer, and the message has the priority, not the sender.

Peer-to-peer connections are not often used in verification systems, because we tend to design systems with a controller in mind. This type of connection is discussed in the next section.

True peer-to-peer connections allow multiple masters and shared communication.

Master-to-slave and push-vs.-pull connections

Contrary to the peer-to-peer connection, most verification components communicate in some sort of unbalanced connection, such as *master-to-slave*. The master initiates an action—either pushing some data to the slave, or demanding/pulling some data from the slave. In either case, the slave must respond.

A *push* connection occurs when one module tells another module to take some data. A common example of this is a generator putting some data into a queue for a BFM to send. The master, in this case the generator, is at a higher layer of abstraction, as opposed to the BFM agent, which "simply" directs the BFM to execute the transaction.

The *pull* form of the master-to-slave connection occurs when one module calls another to get some data. For example, a generator might need to combine data from several sources to form a complete data packet. A specific example of this is when several logical channels share a physical interface. Both the UTOPIA[1] interface and USB interface use this approach.

The appropriate choice of push versus pull is situation dependent. If a connection seems awkward, often reversing the direction of the connection simplifies the code a lot. The general rule is to minimize the number of connections, as well as the assumptions about the connections. If you are uncertain about which type to use, bias your design towards push connections. This is because the decision regarding what to do with the data (for example, whether to send data through a chip interface) is often simpler and of a lower abstraction level than the generator of the data.

Note that at the monitor level, the push is from the monitor towards the checker, because the monitor does not need to know about the recovered data's eventual use. Again, the idea is to minimize the assumptions about an interface.

The following sections are all generalizations of the push/pull interconnection technique.

Most class connections are either push or pull. If the code seems clumsy, try reversing the direction of the connection.

---

[1] Universal Test and Operations Physical Interface for ATM.

Two Tight Connection Techniques

The tighter the connection, the simpler the connection tends to be. This is because the techniques for these connections usually name the class or method that is to be used. This is appropriate for a large number of the connections in a verification system. Let's look at two of the most common techniques: using pointers and inheritance.

Using pointers

Pointers in a project are as common as ones and zeros. So why talk about them? In SystemVerilog, pointers *per se* are not explicitly used. Rather, class instances themselves function as pointers, and therefore can contain data or connect different classes. Specifically, in certain cases it's more reasonable to have two classes implement a task than one. This is common when you are crossing abstraction layers. In one case, one side of the connection has some data but does not want to know how the data will be used. In another case, one side of a layer needs more information to complete a task. In both these cases, it might be reasonable to express the "other half" of a task as a separate class. To clarify this, let's look at a specific example.

Suppose a monitor can gather some data, but because we want to separate the data gathered from any processing of the data, we decide to use a pointer to another class. Here, the monitor could take in a pointer to the other class to "handle," or manipulate, the data, as shown below:

```
class handler;
  extern virtual task check (int data);
endclass
class child_monitor;
  function new (handler p);
    handler_ = p;
  endfunction
  task data_received (int d);
    handler_.check (d);
  endtask
    local handler handler_;
endclass
```

As we can see, using a pointer to another class minimizes the assumptions regarding what happens to the data.

Consider using pointers when abstraction layers must be crossed. Be careful of what each side of the layer "knows."

Using inheritance

Another way to pass the results of one function to another is to use *inheritance*. While inheritance is a very tight form of connection, it can be very clean and provide good separation between roles and responsibilities. Inheritance can be used as the initial connection, while a looser technique can be used to complete the connection. (This two-step approach to connections is used in almost all the examples in Part IV of this handbook.)

But let's look more closely at the concept of using inheritance for connections.

As an example, consider a base class that contains a verification component's algorithms that either consume or generate data. With this technique there would be pure virtual methods to create or use the data; an inherited class would be responsible for providing or consuming the data.

Let's look at how you might create a generic `checker` base class and how you could use inheritance to make the connection to a "real" checker.

```
virtual class checker;
  task start ();
    fork begin
      teal::vout log = new ("Checker");
      forever begin
          data expected = next_expected_data();
          data actual = next_actual_data();
          if (!expected.equal (actual))
              log.error ("Expected: %s != Actual %s",
                    expected.sreport(), actual.sreport()));
      end
    end
    join_none
  endtask
```



```
  'PURE virtual task next_expected (out data d);
  'PURE virtual task next_actual (out data d);
endclass
```

Now you can define a "real" checker:

```
class uart_checker extends checker;
  function new (uart_generator g, uart_monitor m);
    super.new ();
    generator_ = g;
    monitor_ = m;
  endfunction
  virtual task next_expected (out data d);
    d = generator_.next ();
  endtask
  virtual task data next_actual (out data d);
    d = monitor_.wait_for_next_data ();
  endtask
  local uart_generator generator_;
  local uart_monitor monitor_;
endclass
```

Note that the base class, `checker`, has no knowledge of how the data are gathered. This type of connection can be good for separating the responsibility of getting the data from the responsibility of checking the data.

Inheritance is well-suited for the initial connection to the classes outside the base class. In this example, the `uart_checker` is connected to both a generator and a monitor and is waiting for the data. A different implementation class could connect to queues of data. Still another implementation class could use the generator, but for the monitor it would wait for an event and then read the received data from the HDL.

The same two-step technique could be used to minimize the connection assumptions that need to be made for a BFM or monitor. A pure virtual method could be used to consume the data. One subclass might put the data into a channel, while another might filter for special packet processing and then send the data to a specific checker on the basis of this processing. The XON/XOFF processing of the UART interface is a good candidate for this type of connection, as is the processing of Ethernet multicast packets.

> *Inheritance may be a way to defer the specific interconnection mechanism and thus be useful— or it can add complexity, if only one type of interconnection subclass is used in practice.*

Threads and Connections

As we discussed in the overview, verification systems use threads in proportion to the concurrent activities in the chip. Therefore, it's natural to build verification systems that mirror this parallelism. To make the connection between the independent threads, we need a connection that can pass data between threads. This is called a *thread-safe connection*. We'll talk about the base thread safety mechanism, the event, and then move on to fancier thread-safe connection techniques.

Events—explicit blocking interconnects

We now shift our focus from the general types of interconnect to those that can cross a thread boundary. This is important because we use threads often in verification.

Most threads are synchronized by an underlying "wait and signal" mechanism. This is done by an object called an *event*. An event blocks a calling thread until another thread signals that the event has occurred. This technique is the building block of most higher-level interconnect mechanisms, but it can be useful as a technique by itself.

One thread waits for an event to be signaled, while another thread signals the event. This is a good mechanism for coordination, because the waiting thread needs to know only the name of that event. Note that the signaling of the event generally indicates that the other thread has entered a desired state being waited on.

For example, consider a protocol error generator. This error generator forces the wires of an interface to an illegal value during a specific phase of the wire protocol. In order to achieve this, the error generator needs to know the phase of the protocol. It can do this by explicitly monitoring the wires, but it is better to separate roles and responsibilities by using a separate monitor. The monitor is responsible for providing events that trigger on the beginning of the different parts of the protocol. The generator just has to decide what part of the protocol to corrupt, and when, then wait for the specific monitor event. After the event is signaled, it can force the wires into an illegal state until the next protocol state is signaled.

This is shown in the example below, which creates errors in the Cyclic Redundancy Check (CRC) phase of the protocol.

```
class protocol_monitor;
  teal::latch¹ crc_phase_begin;  //CRC is detected
endclass
class crc_corruptor;
  function new (protocol_monitor pm);
    protocol_monitor_ = pm;
  endfunction
  task start ();
    forever begin
      protocol_monitor_.crc_phase_begin.pause ();
      //Now force the wires to corrupt the CRC
    end
  endtask
  local protocol_monitor protocol_monitor_;
endclass
```

After the `crc_corruptor` is hooked up to the `protocol_monitor`, the `crc_corruptor` is started. The latter then waits for the signal from the monitor and then trashes the CRC. We have separated the protocol specifics from the desired action.

Note that there are no data other than the fact that the event occurred. The fact that the event occurred is all that is needed in this example. However, this can be limiting, as threads often exchange data on the basis of some coordinating event. This issue is solved in the following sections.

> *Events are useful as a form of a connection, because the data exchange is minimal. Make sure that the triggering of the event is all that is really needed.*

[1.] A `teal::latch` is a slightly fancy event. It is described in the Teal Basics chapter.

Hiding the thread block in a method

Instead of a just using an explicit event for the connection, consider hiding the event behind a class method. This is called a *blocking method*, because the method blocks the calling thread. By using this technique you can associate data with the event. Also, the event is now abstracted into a method.

Moving to a blocking call makes coding sense, because the choice of using an event, an HDL wire, or a set of events is now up to the implementor of the class. It also simplifies the interface, because method calls are a standard way of communicating. The fact that the call is blocking can almost be an implementation detail. This can make the code clearer or more confusing, depending on whether a user of the class can reasonably expect that the code will block. Sometimes this blocking can be implied by the method name.

As an example, assume that a protocol monitor has the method `wait_for_start_of_frame()`. Because of the "`wait_for`" in the name, one can assume it will block the calling thread. Now the monitor is free to implement the method in any way that best fits a specific design. Perhaps it has an internal event called `start_of_frame_event_` that is triggered by an internal thread. An alternative implementation might have an internal bit variable `start_of_frame_`, and poll it on the positive edge of a clock. Still another implementation might be an internal state machine and a single event that indicates a change in the state. The point is to separate the interface from the implementation, minimizing the implementation assumptions.

Another variant on the blocking method is to use an overloaded method, commonly called `pause()` or `trap()`. There will be several `pause_<name>()` methods, each with a different pointer to an object that specifies the event desired and the data to be returned.

Continuing with our monitor example, suppose the monitor supported three blocking methods: waiting for start of frame, waiting for start of data, and waiting for completion of the data packet.

 • • • • • • •

The following is an example code interface:

```
class start_of_frame;
  uint32 frame_number;
endclass
class start_of_data; //just a class, no data needed
endclass
class data_complete;
  uint32 data[];
  uint16 crc_16;
endclass
class monitor;
  extern task wait_frame (output start_of_frame sof);
  extern task wait_data (output start_of_data sod);
  extern task wait_data_complete
    (output data_complete dc);
endclass
```

This method has the advantage of using a naming convention to show that the methods are related and that blocking semantics are used.

Using a blocking method is often better than using an explicit event. Make sure the method name conveys the block, if the block is not just an implementation detail.

Fancier Connections

The connection techniques discussed above provide a good basis for fancier, more complicated connections. So why did coders invent these fancier connections? The techniques discussed below are combinations of the basic ones, and are used to express the coder's intent better. Although you might not use all these techniques all the time, it is good to have them in your bag of tricks.

Listener or callback connections

Sometimes you do not need a two-way connection, but just need to "listen in" on another object. A technique for this type of connection is called the *listener*, and is sometimes also called the *callback*.

Why two names for the same thing? Programmers often used the term "listener" when they are viewing the architecture from outside of the class to be listened to. Programmers use the term "callback" when they are speaking from the other side, from the class with the interesting data. Confused? Don't worry, this is not a technique that we recommend or use often, as we explain below. We'll use the term "listener" throughout this section.

Listeners are objects that are called at specific points in another object's methods. Often the two objects are unrelated, although often a pointer to the calling object (the one with the interesting data) is passed.

Remember our monitor example, which had a "start_of_frame," "start_of_data," and "data_completed" thread synchronization points? Instead of an object representing just the interesting synchronization points, there could be generic listener objects that would be called at many points in the monitor's state machine. An interface could be like the following:

```
typedef class monitor;
virtual class action;
  'PURE virtual do_action (monitor m);
  //perform action, with monitor state as needed.
endclass
class monitor;
  uint32 current_frame;
  extern task add_start_of_frame_listener (action a);
  data current_data;
  extern task add_start_of_data_listener (action a);
  extern task add_data_complete_listener (action a);
endclass
```

Note that this is functionally equivalent at a high level to what we had in the previous section. However, in this case, instead of a blocking method, an object's do_action() is called. This may make the intended task easier or harder, depending on the task.

If the listener's task is relatively self-contained, as with incrementing a counter, this technique is straightforward. If, instead, the task is to implement some high-level algorithm and that code `case`'d on several of these state changes, the multiple listeners needed would be an extremely clumsy way to express the algorithm.

This technique can also be used when the author of the original code cannot allow an inheritance-based interconnect, or you cannot get access to the source code.

There are several variants of this technique. The listeners could each have their own class, be separated into pre- and postmethod listeners, or act as filters for some data.

As a rule, use the listener/callback interconnect only when you are relatively sure where to put the callbacks. In addition, ensure that there are many simple, loosely connected actions.

Channel connections

A *channel* is a connection technique that manages a queue of data between two or more objects. It is a fancy way to pass the data between classes. The technique is a loose form of connection, because both sides interact through an intermediary.

Channels usually handle the crossing of thread boundaries. One verification component places data into the channel, while another component—possibly at a later simulation time and in a different thread—consumes the data. One side of the connection has no knowledge of or assumptions about the other.

While channels can complicate debugging, there are many situations where they are a necessary trade-off. For example, a generator usually has to send the data to both a BFM and a checker. This can be accomplished by having a channel replicate its data into two channels. As far as the generator knows, it is sending data into only one channel.

Another example occurs when two or more verification components want to send data to a third component, such as a transactor. You can create a channel that takes input from any number of channels and merges the data into a single channel.

Note that a common channel behavior is that the thread consuming the data waits for the data to become available. In this case the channel implementation uses an event.

A channel is specific to the data it contains. In OOP terminology, a channel is a *container class*. Container classes are good candidates for templating.[1]

Channel connections are very useful in verification. They are a loose, thread-safe mechanism for connecting a number of components together.

Action object connections

Sometimes just having a channel with data is not sufficient for what you want to express. In this case you need to have an object that can "do" something. Maybe the object makes some configuration calls to a BFM, or it just sends a burst of data. In any event, this more active connection is called an *action object connection*.

This method of connection combines the channel and the listener techniques. Even though the data in the channel connection are objects, the idea here is that the channel generally contains passive data. With active object connections the channel can contain both control and data. This can lead to code that is obscured and hard to find, and therefore hard to reason about.

With the use of action object connections, there is a channel or queue of objects, each with a listener type of action object, and a single method: `do_action()`. Various objects create these action objects and place them in a command queue. The owner of the queue pops the action object off and calls its `do_action()` method. This method usually calls some configuration methods in the target object and probably also a `put()` or `get()` method. In this way arbitrary sequences of control and data can be queued.

[1.] Be careful with templating, as all vendors currently support a subset of the SystemVerilog language.

Action object connections can be used to synchronize control and data—a good thing when you want to encode configuration settings, and have a generator create sequences of configurations and data using those configurations. However, although a large number of chips can accept configuration changes with data flowing, make sure that capability is intended to be used in production software.

The authors have rarely used the action-object connection technique, but it has proved useful for complex sequencing problems.

Using action object connections in a channel for the complex sequencing of a chip may be appropriate for testing a CPU or graphics chip, but it is probably overly complex for testing an interface or feature.

Summary

In this chapter we have explored various ways to connect classes. Which technique should you use? It all depends on the problem you are trying to solve. Different connection techniques have different trade-offs. In general, try to use the tightest connection you can, because that will be the most obvious connection.

We talked about the two most common forms of connection, pointers and inheritance. Note that the pointer is an instance-based concept, while inheritance is a class-based one. The significance of this is that the class-based technique is static, and thus slightly simpler.

Using events is a good technique to let one thread know when another thread has changed state. Events are fundamental to thread-safe connections.

You may decide to hide the event by using a blocking method call. This is a good technique for loose connections, such as between monitors and error injectors.

Not surprisingly, the most common connection technique (besides pointers) is to pass queues of data. A channel, a common implementation of this technique, crosses thread boundaries and separates the producer of some data from the consumer. It also allows for clever techniques, such as replicating the data to another channel and filtering the data to add errors.

The final technique we looked at was called active object connections. These are used when you need to mix control and configuration with the data. As with listeners or callbacks, this approach can be a slippery slope. Although everything can be expressed in a mixed control and data channel, just make sure you use only what is needed.

For Further Reading

- On the topic of connections between levels of abstraction and within a level of abstraction, *Software Engineering: A Practitioner's Approach,* by Roger S. Pressman, has a several relevant sections. The fancy term for the connections is called "cohesion and coupling."

- The authors are aware of several books and papers on different connection techniques. None are landmark or stand out as "the best" way. As with learning SystemVerilog, this is more of an impedance matching issue, with some books and papers better matched to your learning style and experience level than others.

Coding OOP

Beauty is in the eye of the beholder.

Common paraphrasing of Plato

Coding is a personal endeavor. For many of us it's similar to creating art, and as with any art, there are many styles—some loved, others detested. Why is this relevant to coding? Well, because unlike the case with art, our code cannot stand alone. We are in the interesting position of creating art that, by definition, must work in a community.

Not many engineers intend to create complicated, stand-alone code. This chapter shows techniques, tricks, and idioms that you can use to communicate your intent. When your code is clear and transparent, other engineers can more easily understand, and appreciate, your intent. Code that is appreciated is more likely to be used appropriately, adapted, and, most important, integrated well with the rest of the system.

Overview

This chapter shows some of the coding techniques we can use to create our art. This, the last of the OOP chapters, talks about the coding going on inside a class. Of course, what's going on in a class is related to the class structures and interconnects around the code, so we will not limit our discussions to the lines of code in a method. Rather, we focus in this chapter on coding.

Our first focus is on "if" tests, with a discussion on why this necessary coding construct complicates the code. We'll show some ways to minimize these "if" tests.

We then discuss ways to get your point across, using coding tricks and idioms. We also look at the touchy subject of coding conventions, and try to point out where they help and where they hinder.

The previous part of this handbook used many different techniques, so we figured that one more look, concentrating on when to use individual techniques, might be useful.

"If" Tests—A Necessary Evil

The fewer "if" tests a segment of code has, the easier the code is to understand. Code that just does step one, followed by step two, and so on is inherently easier to reason about. An "if" test, by contrast, causes us to think some more. Is the condition true? Will the code set something up that I have to remember? Where was this condition set?

This section looks at ways to minimize "if" tests in your code. Of course, there will always be "if" tests in code; the goal is to find the place to put them so that their presence is reasonable and maybe even expected.

To put this more generally, writing code consists of procedural statements and changes in control. The mix of these components two dramatically affects the complexity of the code and its adaptability. Procedural statements are just unconditional expressions followed by a " ; "—for example, in function calls or mathematical statements. Because the processor always executes the statements in order, it is not difficult to understand the control flow.

By contrast, changes in control increase the "mental state" that must be remembered. One must now consider two paths through a block of code. These changes in control can be looping constructs, "if" tests, "switch" statements, or "?" operators. This section is primarily concerned about "if" tests. Loop constructs can be considered as combinations of an "if" test. "Switch" statements can be considered as restricted implementations of "if" tests, so they are simpler, but still not as simple as procedural code.

The question mark operator is somewhat like an "if" statement, but, in practice, is best relegated to assignment statements, such as in a = (b > 4) ? 10 : 62. In this case, because it is a common idiom, the operator does not increase the mental complexity of the code.

"If" tests, while necessary, can complicate the code. Using them where they make sense is tricky.

"If" tests and abstraction levels

Almost all algorithms have "if" tests as part of their definition. Algorithms at the BFM/monitor level are probably best implemented with these "if" tests. However, as the abstraction level increases, the number of "if" tests should decrease. This is because "if" tests make it harder to reason about the code. At the higher levels, the code should be fairly straightforward.

If there are "if" tests at the higher levels, they should be more of a "presence or absence" test, as in the following:

```
if (scenario::provide_jtag_traffic ()) begin
   jtag jtag_stimulus = new ();
   ...
end
```

These kinds of "if" tests still raise questions, but questions that should be more easily answered.

At the high level, the "if" test should be rare.

"If" tests and code structure

Be careful about using "if" tests whose body exceeds a few lines. The body is the part of the code that is affected by the "if" test. This can complicate the understanding of the code, because the structure of the algorithm is obscured. The authors have pondered many examples of this.

Sometimes the "if" tests are preventing the last inner code block from executing. A common example of this occurs when you have a series of conditions that must be met (perhaps stages in some protocol) before you execute the inner code. In this case, consider using return statements on the inverse of the conditions to weed out these large blocks of "if" tests.[1]

Let's look at a specific example. Here is a function that has to process a packet of data:

```
task my_class::handle_packet (packet a_packet);
if (a_packet.valid()) begin
  if (i_am_enabled_) begin
    if (we_are_in_sync_) begin
      packet_count += 1;
      if (system::error_level (1)) begin
        //...the algorithm here...
      end
    end
  end
end
```

---

[1] This is yet another case where the academics differ from the industrial coders. You may have been taught "there is only one return from a function," but this rule—like all rules—works best if applied with intelligence, not dogma.

```
    end
  endtask
```

Instead, consider this reorganization of the code with a negation of the original condition:

```
task my_class::handle_packet (packet a_packet);
  if (!a_packet.valid ()) return;
  if (!i_am_enabled_) return;
  packet_count_ += 1;
  if (!system::error_level(1)) return;
  //...the algorithm here...
  end
endtask
```

In this case the algorithm is clearer. There are some preliminary tests, and, if they pass, the main algorithm is performed. This allows you to forget the "bad" cases and concentrate on the core algorithm.

However, this technique of testing is not always appropriate. As with most things in coding, it's a trade-off. The counter-argument against testing for "not true" is three-fold:

- Testing for the negative is sometimes counterintuitive.

- Multiple returns complicate an algorithm.

- Because you do not have indentation in the code syntax to help you, you must remember what tests have passed as you read down the function.

Avoid long "if" tests. Also, sometimes "if" test reversals make the code easier to understand.

Repeated "if" expressions

There is a class of "if" tests that are particularly damaging to reasoning about the code. These are the tests whose expression has already been tested. These tests make us keep rethinking a condition and make us nervous as to why another check is needed. Can the condition have changed from the first test?

In addition, there is a danger that the person reading the code does not see all the repeated expressions and misses some when the expression must be modified.

There are two solutions to this situation, assuming the repeated expression is in a single class or function. One solution is to encode the results of the test into a bit data member and then use this data member instead of the class. The other method is to rethink the algorithm. It is quite possible that the algorithm needs to be "warmed over," recrystallized, or converted from push to pull.

Be extremely cautious about repeated "if" expressions in different objects. This is almost always a design mistake, because the algorithm is being spread out across several classes, making it difficult to change it.

The authors once worked on a piece of code designed to power down the units of a chip that were not being used. However, once we changed the code, the test hung at the end. It turned out that, by powering down the logic block, the chip would not respond to a register read and would hang. In retrospect, this behavior made sense.

The code that was doing the read was confirming that the module had not raised any errors. This was fine, but the test for "module in use" was repeated in another part of the system that the authors had missed.

In the end, this bad style of widely distributed, repeated "if" tests was not all bad for this project. The incorrect "if" test showed that, if software accessed a register of a powered-down module, the chip would hang. The designers ended up adding a simple watchdog timer on the internal bus.

> *Retesting the same condition test is clumsy, complicated, and error-prone.*

"If" tests and factory functions

So "if" tests are dangerous. How do we make as few of them as possible? One way of minimizing "if" tests is to encode the different algorithms in a class hierarchy with a common base class, and use a single function to build these inherited classes. This function, called a *factory function,* is given information to help it decide what inherited class to create. The factory function returns a base class pointer. The rest of the system would not know the specifics of the actual class; for all the code knows, the object is a base class.

This is a very powerful technique for code adaptability. If the factory function is separated from the class hierarchy, it can be adapted to different situations without the need to change the rest of the system, because the rest of the system knows only about the base class.

This reduces the need for "if" tests, because the mechanism of virtual functions is taking their place. Very OOP!

A factory function example

As an example, consider a task to test a Controller Area Network (CAN) protocol implementation. The testbench consisted of a reference model (verification IP), a hardware implementation (which had three ways to drive CAN, DMA, FIFO, and register), and a hardware-assisted peripheral interface controller (PIC) implementation (another small microprocessor on the same chip). Because the CAN protocol is a multinode protocol, several nodes were created. A factory function was used to return a generic can_node class, even though within the factory function one of five possible inherited classes were built.

The traffic generator did not have any knowledge about the specific CAN implementation connected to it. The generator would simply send random traffic. The checker also did not know about the implementation of the nodes. It would just make sure all the nodes were in the same state (after each bit time) and report any differences.

Here is the can_node base class that the generator and checker had access to:

```
virtual class can_node;//The generic base class
   function new (string name, configuration c,
               virtual pins wires);
               //the interface to the chip
     configuration_ = c;
     port_ = wires;
     name_ = name;
   endfunction
   task init ();
     init_();
   endtask
```

```
    task start ();
      start_();
    endtask
    task stop ();
      stop_();
  endtask
  task send_message (message_data m);
    send_message_ (m);
  endtask
    task message_received (message_data m);
      message_received_(m);
    endtask
  function string name();
    return name_;
  endfunction
  'PURE protected virtual task init_();
  'PURE protected virtual task start_();
  'PURE protected virtual task stop_();
  'PURE protected virtual task send_message_
          (message_data m);
  pure protected virtual task message_received_
          (message_data m);
  configuration configuration_;
  virtual pins port_;
  local string name_;
  endclass
```

The base class is the foundation for all the inherited classes. In this
example, it will specify how to start and stop the node, as well as how
to get and send messages. The `init_()`, `start_()`, `stop_()`,
`send_message_()`, and `receive_message_()` methods are pure virtual
methods, which means that the inherited classes must provide their
implementation. This makes sense, because in our example we have
several implementations. The base class would not know how to interact
with the chip.

The `local` section of the base class has only the name of the instance,
and there is no inherent reason that this should be hidden from the
inherited classes. Because the name is a constant, the inherited classes
cannot change it after construction. Note that one may want to append
`_fifo`, `_dma`, `_pic`, or `_vip` to the name, but that should be done in the
constructor of the inherited classes.

Now that the base class is defined, the inherited classes, such as `can_fifo`, can be defined. They are prototyped here, but the implementation is too specific to be explained in this handbook.

A factory function to build the CAN nodes was used. First, here are the inherited classes, without their implementations:

```
class can_node_dma      extends can_node ...
class can_node_fifo     extends can_node ...
class can_node_register extends can_node ...
class can_node_pic      extends can_node ...
class can_node_vip      extends can_node ...
```

Now, an `enum` is defined that can be used to select which specific node type to build:

```
typedef enum {dma, fifo, register, pic, vip} can_type;
```

At this point, assuming that we randomize on what type of node we want to build, a single function can be called to create the specified type of node. This is shown below:

```
//The Factory Function -
// Returns 1 out of 5 possible implementations
function can_node build_can_node (can_type the_can_type,
    string name, configuration c,
    virtual pins p);
  case (the_can_type)
    dma: can_node_dma n = new (name, c, p); return n;
    fifo: can_node_fifo n = new (name, c, p); return n;
    register: can_node_register c = new (name,
                                    c, p); return n;
    pic: can_node_pic n = new (name, c, p); return n;
    vip: can_node_vip n = new (name, c, p); return n;
    default: truss_assert (0);
  endcase
endfunction
```

Given the above factory function, the top layer of the CAN test just builds the nodes and connects them to generators. Because the chip and the PIC were being developed at the same time, we also had separate tests that built only one of those types and a reference node. It was only the top layer of the CAN test that had any knowledge of the specific types of nodes that were being connected. This was later randomized to test various combinations of nodes.

Using factory functions to build a specific inherited class is an OOP technique to reduce "if" tests.

Coding Tricks

When we code we tend to use patterns that have helped us in the past. This section presents a few such patterns. This section, as does the following section on idioms, shows conventions that have helped coders focus their thoughts and tighten their code.

Coding only what you need to know

This is perhaps the cardinal rule for creating good code. It's based, of course, on the same assumption that creates our profession of verification:

Code that is not verified will contain bugs.

Sure, by using this technique we write code that does not have all the features that every situation needs—but a smaller system is easier to reason about and thus adapt. Remember, if a coder needs to add code to your class, it's because they need it.

Another reason for coding only what is needed is because then the code that exists is at least verified to some level. It's frustrating to work on a method, only to find out after hours of debugging that it was never used and does not work.

If a feature is present in some code, it had better be working code. Be cautious about implementing features you will not use today.

Reservable resources

The majority of hardware has a bus to access configuration registers. As the verification system we write consists of many threads, there is a danger that two threads can start using the bus at the same time. The hardware bus is considered a *reservable resource*, because code must first request access to the resource. The trick is to make the reservation as simple as possible for the code that uses the resource.

The simplest solution is to hide the reservation inside the implementation of the class. This is appropriate in most cases. A simple mutual exclusion (mutex) algorithm can be used for this purpose. A mutex only allows one thread at a time into a section of code. The latter parts of this handbook show examples of using a mutex in a register-access BFM.

Sometimes the hardware can process several requests at the same time. This is probably an implementation detail when used in a full chip test, but it is probably something you need to expose at the unit level test. In this case there are two classes. One class exposes a *key*, such as an integer tag or an instance in the interface. The other class uses this lower-level class and hides the key from the rest of the system.

The concept of reservable resources can also exist solely in the verification system itself. You might, for example, have a DMA descriptor queue and need to allocate and release descriptors. Of course, the hardware actually implements the queue, but the management of the queue is a verification concept.

> *Reservable resources may be an implementation detail, and thus use a mutex internally—or they may be an external property, in which case a "key" must be used. This key can be anything from an integer to an object, depending on how safe the management must be.*

The register: an int by any other name

Accessing a chip's registers is an important part of a verification system. Register access consists of three parts: the register's address, data, and fields. The authors suggest that none of these be classes. Why? Because in production software (which has only memory access and interrupts), the register address will be an integer, the register itself will most likely be an integer, and the field names will probably be macros.

Consider register fields. It is often a good idea to assign register fields by using a field name, rather than hard-coded integer offsets. This makes the code clearer and allows fields to be relocated within the register with little pain. In Part IV of this handbook, we show some simple `'field_get()` and `'field_put()` macros.

When considering how to write and read registers, the authors prefer to use an indirect, but simple, technique. We use the Teal memory functions, which take in an integer address and a bit array for the data. The actual protocol used is then appropriately simple (and abstract) for the test writer. We use the testbench to attach a protocol to a specific address range.

As another example of the utility of this approach, a verification system may have back- and front-door register reads, and choose which to use based on a test parameter. Also, there may be multiple front-door implementations, such as through a processor bus, a PCI, or even a JTAG[1] protocol. One of these techniques could be selected randomly.

> *Registers are your friend, but don't use them as exercises in OOP. Keep it simple.*

Using data members carefully

When you start building a class, there is a tendency to make many data members. It is common to see a number of calls that have no parameters, but that use the data members in the class as a shorthand. This is fine when those methods are called from outside of the class. However, for

---

[1] Joint Test Action Group, IEEE 1149.1

a protected or local method that is called by a public method, consider using the standard parameter passing instead of a data member.

Here is an example:

```
class a_class;
  local int value_;
  local int weak_data_member_;
  //called from outside the class, use data member
  task method1 (int value); value_ = value; endtask
  //called from outside the class, use data member
  task method2 (); value += 1; endtask
  task method3 ();
    //Is this confusing?
    weak_data_member_ = value_ + 3;
    method4();
    method5();
  endtask
  local task method4 ();
    weak_data_member_ + = 10;
  endtask
  local method5 ();
    value_ = weak_data_member_;
  endtask
endclass
```

The reasoning is that a data member is a bit like global state and comes into a method whether or not you want it to. As such, it makes the class slightly more complicated. This is fine where it is necessary, but inappropriate if the data could have been simply passed in as a parameter.

The fancy term for all this is *spatial locality*. In our case this means that the data are needed by multiple calls *from outside the class*.

A related fancy term is *temporal locality*. This refers to code that is in different functions but is *called sequentially*, as follows:

```
begin
  object1.do_method();
  object2.do_another_method();
end
```

In general, with spatial locality you want to use data members. With temporal locality, you want to use parameters to the calls.

Here is the example reworked to pass parameters (this example has temporal locality, but not spatial locality):[1]

```
class a_class;
  local int value_;
  //other methods as before...
  //less confusing?
  task method3 ();
    method5 (method4 (value_ + 3));
  endtask
  local function int method4 (int temp);
    return (temp + 10);
  endfunction
  task method5 (int temp);
    value_ = temp;
  endtask
endclass
```

Use data members sparingly. Make sure a data member is needed because of spatial locality.

Coding Idioms

An *idiom* is a fancy word for a coding trick that can be expressed not only as a concept, but also in a well-known code structure. This section introduces some idioms that the authors have found to be useful for building verification systems.

[1.] Yes, this is from a real test system. The authors have changed the method names to protect the original coder.

The singleton idiom

Sometimes a class is meant to be instantiated only once, and it has no clear owners. The fancy term for this is *global service*, as was discussed a bit in the chapter on OOP classes. Let's look in detail at a common implementation of this one-off instantiation, the *singleton*. A singleton uses a single static method, called `get()`, to return a pointer to this single instance.

Consider the following example:

```
class channel_counter;
  static function channel_counter get ();
    assert (channel_counter_ != null);
    return channel_counter_;
  endfunction
  static task start ();
    assert (channel_counter_ == null);
    channel_counter_ = new ();
  endtask
  static task stop();
    channel_counter_ = null;
  endtask
  local static channel_counter channel_counter_;
endclass
```

Another common convention for singletons is just to have a global function that returns a pointer to the global object. This global function may be put into a package if it makes the idiom clearer.

Note that the creation of the internal implementation pointer is a different matter. There are different ways to do this, from automatically creating one on first use, to having a factory function, to having an static `start()` method. Which mechanism to use is a personal choice.

Singletons are a good way to implement a global service.

Public nonvirtual methods: Virtual protected methods

When you are coding a class, there are often *virtual functions*. These methods provide the implementation of either the whole interface of the class, or perhaps just a few specific details. Your first instinct is to make these virtual methods public, and this might be good. However, sometimes you need to do some basics things first, or perhaps afterwards. How do you guarantee that the pre- or postcode is called?

The trick is to have a *public nonvirtual method* that just does the pre- or postcode and then calls the same named method (with an identifier, such as a trailing underscore) as a *virtual protected method*. This allows any standard preamble or postamble code to be guaranteed to be executed. Sometimes you might want to use this trick even if there is no special code. It's a useful technique to separate an interface method (those with an underscore, or "_") from an implementation method.

Here is a short code snippet:

```
class my_thread;
  task start ();
    start_(); thread_count_++;
  endtask
  protected virtual void start_();
  local uint32 thread_count_;
endclass
```

In this case the public code interface is through the `start()` method. The actual implementation is done through inherited classes by means of the `start_()` method. This allows a reader of a class to concentrate on the public, "nonunderscored" methods. It also allows coders that need to inherit from this class to concentrate on the protected, "underscored" methods.

With this technique, the nonvirtual public method is firmly in control and calls the virtual method only after performing any desired pre or post actions. Sometimes, though, the very nature of the call expects pre- and postconditions. In this case it is clumsy for the inherited class to have to remember to call the base class method. If the designer of the base class wants to encourage, or anticipates, such usage, it's better to add virtual pre and post methods explicitly.

Here's a code snippet that can be used in this case:

```
virtual class generator;
   task generate_one ();
     __generate_one ();
     //code here to do the standard generate_one()
     packet_count++;
     generate_one__ ();
   endtask
   'PURE protected virtual void __generate_one();
   'PURE protected virtual void generate_one__();
   local uint32 packet_count;
endclass
```

In this case the "main" method—generate_one()—is not virtual, but the pre and post methods are. One convention the authors have used is to write pre_ and post_ as prefixes to identify the set of methods. However, the convention that the authors prefer is to name the pre method the same as the main method, but with a double underscore ("__") prepended. The post method is similar, but with a double underscore appended. In this convention, the reason the letters "pre_" and "post_" are not used is that they can interfere with the semantics of the name of the original method (which might be something like post_process, or post_completions, or prefetch_data). As is a common theme in this handbook, the choice is yours[1].

To enforce the calling of special pre or post code, use combinations of public nonvirtual methods and protected virtual methods.

---

[1] Of course, pre_randomize() and post_randomize() are reserved methods in SystemVerilog.

Enumeration for Data, Integer for Code Interface

Enumerations (enums) were introduced in programming languages to make the code clearer. They are more powerful than defines.

Using enums when setting up parameters can increase the communication level, but there are a few dangers. One occurs when the enum is "`case`"d, or "if" tested. This is can lead to unexpected behavior when enumerations are added. *Enums should generally be used as a shorthand for integral values.*

To this end, the method that uses the enum should sometimes take in an integer as the formal parameter. Why? Because this allows for future expansion (enums cannot be subclassed) as the integral value of the enum becomes the important part of the method's implementation.

For example, consider a baud rate enumeration:

```
package uart_configuration;
  typedef enum {b_9600 = 9600, b_19200 = 19200,
                b_921600 = 921600} baud_rate;
endpackage
class uart;
  extern task new_baud_rate (uint32 new_value);
  local uint32 baud_rate_;
endclass
```

Again, this is one of those things that you were probably not taught in class. You would have been told to define an enum and use it in all parameter declarations. That technique does work a fair amount of the time, particularly if the range of the enums is fixed for all time. However, in the messy world of coding for a living, sometimes we need to be a little more flexible.

Sometimes you should define an `enum` in a package, but take in an `int` for the methods that would have used the `enum`. Note that mixing enumerations and integers is not always desirable, as it weakens the abstraction. The idea is to use this technique only when future derivations need it.

What's in a Name?

For some reason, class names in OOP tend to be more important than structure names in C or modules in Verilog. Maybe this is because in OOP coding, we can enforce what operations are allowed in a class, so we tend to pay more attention to their names. At any rate, this section provides guidance on how to make the transitions between file, class, and instance when finding your way around a verification system. As we have said many times in this handbook, it's up to you and your team to decide what conventions to use.

Keeping class name the same as file name

A common convention is to have the class name be the same as the name of the header file that declares the class. For example, it is much easier to find a class or definition by using the Unix `find` command directly, rather than piping it to `grep`.[1]

A corollary convention is to have only one class declaration per file. However, there are a few exceptions to this guideline. One is when there are small utility objects that are used only right where the main class is used. Another exception is when you are writing VIPs and it is simpler for a user to understand the interface as a monolithic entity. Note that in some cases, the monolithic header file may just contain an `'include` of other header files.

Consider having a one-to-one relationship between class and file. Exceptions are where there are tiny helper classes and when a group of classes is more important than the individual classes.

Keeping class and instance names related

While you can use any identifier for a typename and a variable, strive to keep names as similar as possible. This seems like an obvious guideline,

---

[1] What could be easier than this? `find <path> -name "*.sv" -exec egrep -l -i "your search text" {} \;`

but we programmers are a lazy bunch. It is simple to miss changing an instance name when a class or enum is changed. It takes work and typing to keep names simple. (Appreciate that we essentially type for a living.)

Consequently, when an instance of a class is created, try to name the instance the same as the class. Sometimes, if there are several instances of a class in the same scope, a "_<n>", where n is an alphanumeric variable, can be appended to the name. The reason is that this provides a good mental link to back to the class definition, which specifies what can be done with this instance. A counter-example is when a class provides some generic behavior that can be used in many contexts. For example, a register class may provide generic reads and writes, as well as take in an address in the constructor. In this case it is the mnemonic of the address that is the best name for the instance.

Here is another counter-example, from a project the authors worked on:

```
ht_vip ht_drv; //hard to remember that ht_drv is a ht_vip
//Is pex known in the project? Is mon better than monitor?
pex_mon a_pex_mon;
```

Note that SystemVerilog allows identifiers and typenames to have the same string, as in `my_class my_class`, but the identifier shadows the class in some contexts and can be confusing.

Instance names should be readily traceable to their class name.

Coding with Style

Coding conventions can quickly become a "religious war," something that is not productive for a project, team, or individual. As a remedy, this section presents some style conventions that have proved to be useful. However, as with all the other sections in this handbook, the recommendations made are not intended to be a set of rules.

Adhering to a single style may improve clarity, but only if the entire system is coded by a single person (think "My style is the best!") But even in this case, one's style often evolves over time and adapts to the style of a team. In the general case, the industry definition of "good style" evolves as well.

Because of the evolution of what makes "good style," differences in style are essential for the learning process.

Proceeding with caution

In general, coding conventions slow down good coders, and do not necessarily increase the readability of the code created by poor coders.[1] Understandable code is understandable code, independent of the conventions used. The goal of a coding convention should be to increase communication among the team members.

For teams that feel the need for a "team style," a "guidebook" is usually a better idea than a required coding style. This guidebook should include guidelines, with reasoning following each guideline. In addition, counter-arguments where the guideline may not be appropriate should also be provided. If the entire team does not agree on some guidelines, it is a good idea to include both the pro and the con arguments, so that locally appropriate decisions can be made—including allowing each team member to decide. The presence of the counter-argument also provides a framework should some assumptions change.

The goal of a coding convention should be to increase communication among the team members, not slow down the fast coders.

General syntax conventions

One guideline is to use all lowercase identifiers, with underscores and separators. Identifiers are all the nonreserved words of the language: your variables, class names, methods, data, and enums. The reasoning behind this convention is that there is less time spent thinking about how to type an identifier. An exception to this guideline would be if the team wants to capitalize three-letter acronyms (TLAs) and macros.

[1.] "All generalizations are false, including this one." — Mark Twain

A counter-argument to this approach is that it can create long names.

An alternative convention uses capitalization to indicate an identifier's scope, or class. For example, method names could begin with a capital letter, while data members begin with a lower-case letter. The reasoning behind this guideline is to encode the type information in the case of the identifiers.

Consistent naming conventions can be useful, but beware of dogma.[1]

Identifying local and protected members

Another convention used is to identify protected and local members (data and methods) by using a trailing underscore. This allows one to know quickly whether the method is "internal." It also allows one to look at an algorithm in a method and separate the "internal state" from the method's parameters.

The counter argument is that a method name may become public as the project evolves. Because SystemVerilog uses access rules first, and then scope rules, you can have an issue with code that compiles one day and later does not compile should a method change its access. This issue can also be present if the name had to be changed. Consequently, changing the name may also cause compile errors.

Identifying local and protected members helps others learn about a class.

---

[1] "A fanatic is one who can't change his mind and won't change the subject." (Sir Winston Churchill)

Summary

In this chapter we looked at some of the techniques used to create our code "art." We talked about being careful with "if" tests; they are a necessary evil that can complicate the code. We introduced the concept of the factory function, useful in building inherited classes.

We offered the advice that you should code only what you need to know. We then introduced a variety of useful coding tricks and techniques that experienced coders use to solve programming problems, including, but not limited to, the following:

- Using reservable resources and mutex

- Using register fields instead of hard-coded integers

- Using data members (always carefully!)

- Using idioms to provide structure

- Using singletons for global services

- Using virtual protected methods, to separate code interface from implementation

- Using naming and coding conventions to express intent and understandability

We also presented reasons for considering additional techniques, such as the following:

- When to use enums or integers, and when you should mix them

- Why coherent class naming is a good thing, and why the names of classes, files, and instances should be related

- Why consistent style and syntax are a good thing—if they are applied with intelligence

So, this chapter covered a large number of techniques. Remember, you don't have to use all of these tricks all of the time, but they are here for reference when you need them.

For Further Reading

For a list of resources applicable to this chapter, just revisit the For Further Reading section of the Why SystemVerilog? chapter.

Part IV: Examples (Putting It All Together)

This is what the rest of the book has built up to. Everything discussed earlier in this handbook is applied here to examples that better resemble the real world. This is still a book, so examples need to be relatively simple or they would be incomprehensible, but our goal with these examples is to show what real hardware verification with SystemVerilog looks like.

The examples here build on everything discussed so far. They use the Truss verification methodology, and the Teal classes and functions. They apply the OOP techniques discussed throughout the code.

The examples were not specifically chosen or coded to highlight the strengths of Truss, Teal, or even SystemVerilog. Rather, they were coded to resemble real-life projects as much as possible. Our goal is to show realistic examples and creative solutions. We hope you can pick up an idea or two by reading this. (The code freely available at www.trusster.com also provides a few open-source VIPs that can come in handy.)

Block-Level Testing

C H A P T E R   1 4

I can give you a six-word formula for success:
Think things through—then follow through.

Sir Walter Scott

In many endeavors, follow-through is everything. From sports to parenting, it's not only what you say but what you do that is important. This chapter is the first of the "follow-through" chapters.

We use all the tools, tips, and techniques from the rest of the handbook and apply them to something resembling a real-world example. This is the first complete example of what a test system using SystemVerilog might look like.

We look at a block-level verification environment. Later, we'll adapt this same environment to be used at the full-chip level.

Overview

This chapter covers a block-level verification effort as part of a large project. The goal is to verify a UART 16550 RTL block, written in Verilog. To do this, we will build an environment that will not only verify the block but also provide adaptable verification components for later project stages.

The example presented here will show all verification components needed to do UART 16550 verification as well as a fully randomized test. Several points of interest in the code will be highlighted throughout the chapter. (We present code in a slightly different form from the source code.[1] Sometimes we merge the interface and implementation of a class together, although they are separated in the source code. Also, we may abbreviate a class interface or some method's code to get straight to the point.)

This chapter differs from the Truss tutorial chapter (in Part II) in that it focuses more on the middle layers of a verification system instead of on the flow. The middle layers are where managing the complexity of a verification system comes mostly into play.

If you want to look closer at execution order, it's recommended that you start by referring to the *Truss standard test algorithm* known as the *"dance"* in the Truss Basics chapter. Then, with the "dance" as a reference, divide and conquer by using an ends-in approach. In other words, take a closer look at both the top-level `block_test.sv` (and its related `test_components`) and the protocol aggregrator, `testbench.sv`. This will help show the overall structure and flow of the environment.

This chapter will talk about a few things. First, we set up the example with a theory of operation. That section highlights the overall environment and the protocols that are used.

Then we look at several points of interest in the code. These points cover code complexity problems of the middle layers in a verification system. We present these middle-layer techniques in their order of execution, by first looking at power-on reset, then at configuration and traffic generation, and then at checking.

[1.] Available at www.trusster.com.

Finally, we show how all the pieces are connected together through the `testbench.sv` and `block_test.sv`. This includes details on how channels, configuration objects, and interface layers are instantiated.

Theory of Operation

Many systems have at least one UART connection. This may be for diagnostics, software debugging, or general communication. For this reason, a single UART serves as a good first block-level example.

Here are the main components involved in the simulation:[1]

The UART Verilog core[2] was not developed by the authors. Many implementations of UART cores are available, and it was important to the authors that a known-to-be working UART model be used. For this example we chose an open-source design IP of a UART 16550 from OPENCORES.[3]

This core follows the common register set of the 16550 UART, a popular UART implementation by National Semiconductor. It is so common, in fact, that software drivers for the UART 16550 are included with many Linux distributions. As with all design IP, this core has its own quirks that must be handled. We'll talk about this in the configuration section below.

The UART 16550 core used has two interfaces. One is the actual UART transmit and receive lines, and the other is a local bus to read and write registers in the UART block. In this case the local bus is a *wishbone* interface, a standardized local bus for many OPENCORES models. The wishbone interface will be described in more detail in a later section.

---

[1] Unlike previous illustrations, this one shows the least abstract (most concrete) layers at the top, because now we focus on the concrete layers.

[2] A core is an HDL module that provides a well-bounded functionality. In our case this is the UART16550 registers and UART wire protocol.

[3] See www.opencores.org.

UART Example: Objects and Connections

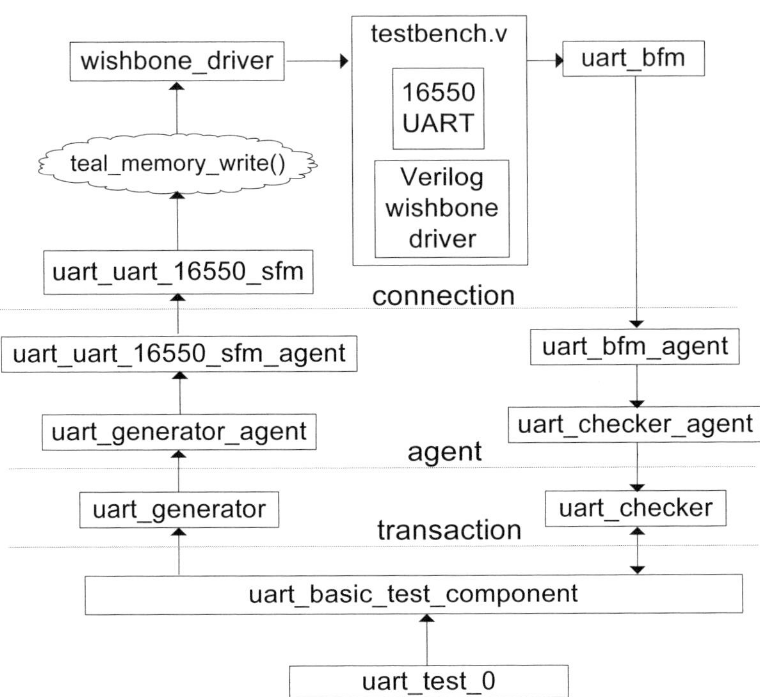

Verification environment

The verification environment uses a UART BFM model to monitor the actual data transmitted and a wishbone driver model to read and write UART registers. For both interfaces, verification IP models, generators, BFMs, configuration objects, and checkers will be designed.

Looking at the Verilog side, the testbench environment is fairly straight-forward. It is shown on the following page.[1]

The UART module under test is called `uart_top`. It's instantiated in the `testbench.v` file, which also instantiates the wishbone Verilog driver and reference clocks. The rest is driven by the testbench. The wishbone driver has a reset wire (not shown), which is used to reset the chip. These are the main components of the system.

---

[1] Here we can actually use the term "wires."

UART Example: HDL Connections

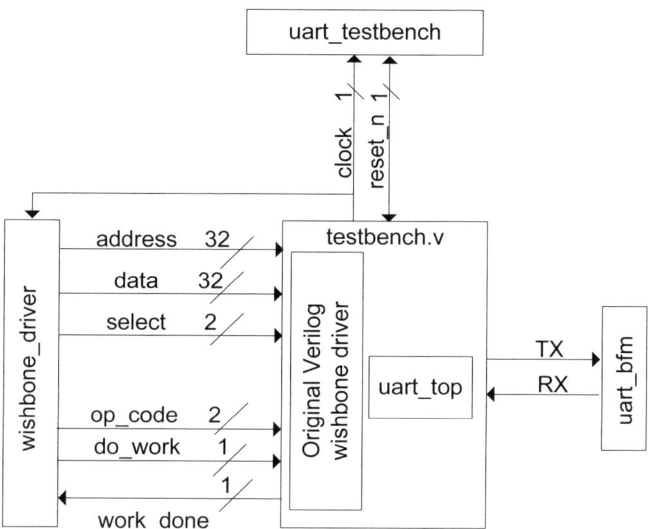

Verification IP

For this example, a Verification IP (VIP) will be adapted or developed for the core's interfaces. VIPs are used to highlight how adaptable verification components can be developed and moved from one project to another. There is always work required when adapting an existing component to a new environment, but if the component is structured appropriately, the work can be minimal.

UART VIPs

For the UART 16550 verification system, the authors developed a generic UART BFM. We also developed what the authors call a *software functional model* (SFM). An SFM is a model of some protocol or common implementation that uses register access instead of wires for the chip connection. An example of this is the USB Open Host Controller Interface (OHCI). The specification for a protocol defines what registers must exist and their meaning. It is similar to a BFM, except that instead of connecting to a bus, we are connecting to registers.

The generic UART BFM is specifically designed to be separate from the specifics of the 16550 protocol, so that the UART BFM can be used for any UART implementation.

The UART 16550 SFM, in turn, deals with all the registers for the 16550 protocol. This SFM acts like a software driver, in that the SFM programs the UART core's registers.

The UART 16550 SFM actually uses the wishbone BFM to set the registers of the Verilog core.

Wishbone VIP

The wishbone protocol is the bus used to read and write the registers of the UART core. For the verification system, a wishbone BFM is used to do this.

Instead of developing a new wishbone BFM from scratch, the authors decided to reuse a Verilog verification model provided with the UART 16550 core. This verification model included Verilog tasks for reading and writing registers over the wishbone interface.

These read and write tasks are wrapped into a BFM. When a Verilog task is to be called (because of a higher-layer testbench call), the testbench side sets the appropriate wires and raises `do_work`. The Verilog side, in an `always` block, then calls the appropriate Verilog task or tasks, and when they return, raises the `work_done`. This signals the testbench side that the results from the Verilog driver are available.

Reusing existing verification models like this highlights how known-to be-working models can be integrated with a new verification environment. This technique is talked about more in the section on reading and writing registers.

The verification dance

The dance is the flow of events (or method calls) during simulation. It, of course, follows the "dance" talked about in the Truss Flow chapter. First, the chip is brought out of reset, then a configuration is chosen by means of randomization, and the UART core is configured (by means of register reads and writes). After this, a generator is asked to generate a group of data words for transmission. Because the UART protocol is bidirectional, both the ingress and egress sides have a generator and checker. After the data have been transmitted, the test waits for the checker to indicate that it received all the traffic. Then the test exits and a final status is printed.

Running the UART Example

Running the example is the same as for all tests that use Truss. However, before you run it, you'll need to set up some environment variables.

In the directory `/examples/single_uart/bin`, there is a setup script. Before you execute the script, make sure you have defined `BOOK_HOME`. Then source the setup script, and it will set `TRUSS_HOME` and `PROJECT_HOME` according to the `BOOK_HOME` variable.

Before you run the `truss` script, you must define the `SIMULATOR_HOME` environment variable. In addition, you must define `SIM`, for the simulator name and path to the install directory you are using. Type `truss -help` to see the currently supported list of simulators.

To run the example, type the following:

```
$TRUSS_HOME/bin/truss --test block_uart
```

There are many other options to `truss`, but this command will compile all the code and run the test. You should see a bunch of compiles and then the test will run.

Points of Interest

There are many points of interest in this first real example. There is a UART configuration object, used to pick the various settings for the protocol. There is a code interface, implemented as a Teal front-door memory bank. There are agent objects, used to connect the UART BFM and SFM to the generators and checkers.

So why are these objects important? In your verification system, you probably will have the equivalent objects for each of these points of interest. Understanding the machinery of their construction and connection—and the trade-offs you may have to make—will help you in building your system.

Configuration

Most chip protocols have a number of registers to support configuration settings. These registers control the exact behavior of the protocol; in other words, they describe what mode it is in. This might determine how the protocol responds to interrupts or whether it uses even or odd parity. With any real chip verification, there is a need to randomize this setup so that each configuration setting is tested.

By creating a *configuration class* for each protocol, you create a centralized place that controls the randomization of each instance. This configurations class is independent of the protocol registers, containing protocol-generic features that are then mapped to registers by a specific implementation.

Why do this? For two reasons. First, you will most likely have a generic side to the protocol, which will operate outside of the chip. This code will not have configuration registers (because it is not hardware), and it can execute the more-generic protocol configuration directly. Second, using a protocol-defined, but generic, class is a way to make the configuration adaptable to other implementations. Moving to a different core or even to a different chip using this same protocol should not radically change the configuration class.

The configuration class is responsible for keeping track of all parameters of an interface, as well as for randomizing them into a "legal" configuration setting.

In our UART 16550 example there are two configuration classes: a generic class for the UART BFM, and a specialized UART 16550 class for the specific UART protocol we are testing. The specialized one inherits from the generic one.

Several techniques are used in the configuration objects to create adaptable code. These, or similar, techniques might be good to consider when you have to write code to verify a protocol. Here we will look closer at the different configuration classes, and highlight interesting areas.

VIP UART package

The VIP provides a generic package that contains the enumerations and associated defines for the configuration UART. This technique of using a package for enums solves the issue that enums are generally in global scope.

```
package uart;
   typedef  enum {none=0, even, odd, mark, space} parity
   //other parameter defines
   parameter int max_uart_width = 32;
   typedef  bit [max_uart_width:0] data_type;
   typedef enum {
            b_150 = 150, b_300 = 300, b_1200 = 1200,
            b_2400 = 2400, b_4800 = 4800, b_9600 = 9600,
            b_19200 = 19200, b_38400 = 38400,
            b_57600 = 57600, b_115200 = 115200,
            b_230400 = 230400, b_921600 = 921600
   } baud_rate;
endpackage;
```

The reason each legal parameter is defined as an enum is to help show intent. This is important when writing adaptable code, and the idea with this UART model is that it should be able to be reasoned about and be adaptable to many different situations.

VIP UART configuration class

The VIP provides a generic UART class that contains the largest legal configuration space. This is because it has been built to be valid for any UART core. The UART 16550 configuration object, by contrast, inherits from this code interface, but actually limits the number of possibilities to what our core can support.

The UART configuration class is described below. It contains an `enum` definition for each parameter that describes the valid protocol domain, as well as a variable for each parameter.

The class also contains `do_randomize()` and `sreport()` methods to set up and print the status of the current setting.

Here are the interesting parts of the class definition:

```
class uart_configuration:
  extern function new (string name);
  uart::parity parity_;
  baud_rate baud_rate_;
  uart::stop_bits stop_bits_;
//other parameter instances
  extern virtual task do_randomize();
  extern virtual function string sreport();
endclass
```

The `do_randomize()` method is responsible for setting each parameter to a legal value so that it can later be written to hardware or interpreted by a generic protocol VIP. Because each parameter is an `enum`, some care must be taken for randomization. Let's look at this.

Randomization of parameters

Randomization is somewhat like logging: it appears simple and obvious, but becomes very complex when a large system is verified. One of the primary decisions is whether to "hook into" the randomization provided natively in SystemVerilog. As you see in the code, we sometimes use Teal's randomization, and sometimes SystemVerilog's. We just want to show what both techniques look like.

In addition to deciding which randomization to use, you have to decide whether a method call—either SystemVerilog's `randomize()` or Truss's `do_randomize()`—should be present in the code interface. Sometimes the fact that there are randomized parameters is just an implementation detail. Other times, however, it is up to the caller to randomize the instance.

In this example, we show how to keep randomization as an implementation detail and use ranges to constrain the randomization. (The next chapter shows an example of how to bring randomization to the interface.)

A utility class, `__uart_configuration_chooser`, is used to do the actual randomization. The dictionary is used to get the min/max bounds and this information is used in the SystemVerilog constraints. The code to choose the parity is shown here:[1]

```
class __uart_configuration_chooser;
  rand parity parity_;
  local parity min_parity;
  local parity max_parity;
  constraint valid_parity {
          parity_ >= min_parity; parity_ <= max_parity; }
  //...
  function new (string n);
    log_ = new (n);
     min_parity = parity' (teal::dictionary_find_integer
                           ({n, "_min_parity"}, 0));
    //...
  endfunction
endclass
```

As can be seen, each time the helper class does the actual randomization. This class is used by the `do_randomize()` method, like this:

```
task uart::configuration::do_randomize();
  __uart_configuration_chooser chooser =
     new (log_.name ());
  'truss_assert (chooser.randomize ());
  parity_ = chooser.parity_;
  //...
endtask
```

---

[1] For the complete code, see `uart_configuration.sv`.

This code shows how the parity is randomly picked by the `do_randomize()` method. This technique of defining a random data member in the utility class and then setting its constraints by means of the dictionary is repeated for each parameter. Thus, at the end of the method, all parameters are set to a random legal value.

UART 16550 configuration class

In this project a UART 16550 core IP is used. The UART 16550 is a common protocol, but our core puts a few restrictions on the legal UART register values. As shown below, we created a valid UART 16550 configuration by expanding upon the generic UART configuration class:

```
class uart_configuration_16550 extends uart_configuration;
  function new (string name);
    super.new (name);
  endfunction
  virtual task do_randomize ();
    //correct cases that our core cannot handle
    super.do_randomize ();
    if ((stop_bits_ == configuration::two) &&
        (data_size_ == 5)) begin
      stop_bits_ = configuration::one_and_one_half;
      log_.debug (
        "Corrected stop bits from 2 down to 1.5
        data_size is 5).");
    end
    if ((stop_bits_ ==configuration::one_and_one_half)
       && (data_size_ >= 6)) begin
      stop_bits_ = configuration::two;
      log_.debug ("Move stop bits from 1.5
          up to 2 (data_size is 6, 7, or 8)." );
    end
  endtask
endclass
```

The `configuration_16550` class inherits from the VIP configuration class. It overrides the `do_randomize()` method of the base `configuration` class. As shown in the implementation of the overloaded method, `configuration_16550` calls the base class method [see the `super.do_randomize ()` line above] and then checks the actual values of a couple of registers.

If, for our core, illegal register combinations have been randomly chosen by the base class, `do_randomize()` corrects it. This is done to ensure that a legal UART 16550 configuration is picked.

Configuring the Chip

So how does an the actual chip get configured once a configuration object has been created and randomized for an interface? The configuration object represents the information a software driver would have to know to set the correct registers in an actual chip.

In the Truss solution we follow this concept in the driver or BFM. A configuration object is known by all the particular drivers, BFMs, and monitors on a protocol. This knowledge is necessary for the connection-layer objects to be able to drive and monitor the physical connections.

But how does the configuration get programmed to the actual chip? This is not normally done over the same protocol. Rather, programming the chip is normally done over one or a couple of major protocols. For example, if a chip has an embedded processor, programming is mainly accomplished through the processor's external address and data wires. If the chip does not have a processor, this is accomplished through some standard, well-defined protocol, such as USB or I^2C.

In our chip the wishbone protocol is used to program the registers in the chip during the write-to-hardware phase of the "dance." The `write_to_hardware()` method of the `uart_16550_bfm` class doesn't access the hardware directly through its own wires. That would both complicate the code and made it harder to adapt. Instead, it uses the register defines on top of Teal's memory routines. The wishbone driver is hooked underneath these memory routines. Let's look at the technique of using Teal's memory access.

Register access

In order to be clear and to create adaptable code, we have the method `uart_16550_bfm::write_to_hardware()` use register writes.

Here is the method:

```
task uart_16550_bfm::write_to_hardware();
  teal::uint8 data;
  data = 0;
  //...
  'truss_assert(configuration_.data_size_ >= 5);
  'truss_assert(configuration_.data_size_ <= 8);
  'field_put(data, data_size,
                  configuration_.data_size_ - 5);
  'field_put(data, access_clock_divide, 1);
  teal::write (UART_REG_LC, data, 8);
  teal::uint8 lc_save = data;
  data = divisor;
  teal::write (UART_REG_DL1, data);
  data = divisor >> 8;
  teal::write (UART_REG_DL2, data);
  'field_put(lc_save, access_clock_divide, 0);
  truss::write(UART_REG_LC, lc_save);
endtask
```

Notice that the code is using both `teal::write()` and `'field_put`. What are these? The `write()` function uses Teal's memory manager to abstract away how the register will be written. The `'field_put` is a local macro that may be useful. It is defined as follows:

```
'define field_put(data,field,value)   \
              data['field"_max:'field"_min] = value
'define field_get(data, field)\
              data['field"_max:'field"_min]
```

Why all this define trickery? The point is to abstract how the actual registers and fields are accessed and manipulated. The authors are aware of, and have created, several fancier ways of accessing registers for verification. However, we believe that this mechanism has the appropriate level of simplicity and opens the door for the software team to understand the verification code.

Notice that the register addresses are defines. This is appropriate, although they could have been `parameter const int` should the team decide that is more appropriate.

The field names are also defines, but they are named a specific way. This is because the `'field_put()` assumes a `_min` and a `_max` suffix to the field names. This was done to minimize the parameters into the macro.

For example, the following is used for the `data_size` field:

```
'define data_size_min 0
'define data_size_max 1
```

Recall that the implementation of `teal::write` will find a memory bank mapped to that address and use it for the actual access.

Next we will look at how an actual address resolved to the wishbone interface.

The wishbone_memory_bank and wishbone_driver

Now we have seen how the UART 16550 SFM writes registers. But how does this get translated into accesses to the wishbone driver? Remember that Teal's memory routines use a look-up table to figure out which `memory_bank` object should handle the memory access.

We'll just add a wishbone memory bank:

```
typedef class wishbone_driver;
class wishbone_memory_bank extends teal::memory_bank;
   extern function new (string n,
                        wishbone_driver driver);
   extern virtual task from_memory(teal::uint64 address,
     output bit [MAX_DATA - 1:0] value, input int size);
   virtual task to_memory (teal::uint64 a,
          bit [MAX_DATA - 1:0] value, input int size);
     wishbone_driver_.to_memory (address, value, size);
   endtask
endclass
```

The real work is done in the wishbone driver, although that, too, just calls down to a module in the Verilog. Here is how the write method of the driver works:

```
task wishbone_driver::to_memory(
    bit [63:0] address,
    input bit [MAX_DATA - 1:0]  value, teal::uint32
        size);
  mutex_.get (1);
  wishbone_driver_interface_.op_code_ <= 0;
  wishbone_driver_interface_.address_ <= address;
  'truss_assert (size <= 8) ;
  //put the data on the right line
  case (a % 4)
    0: begin wishbone_driver_interface_.select_ <= 1;
         wishbone_driver_interface_.data_ <= d; end
    1: begin wishbone_driver_interface_.select_ <= 2;
         wishbone_driver_interface_.data_ <= d << 8; end
    2: begin wishbone_driver_interface_.select_ <= 4;
         wishbone_driver_interface_.data_ <= d << 16; end
    3: begin wishbone_driver_interface_.select_ <= 8;
         wishbone_driver_interface_.data_ <= d << 24; end
  endcase
  wishbone_driver_interface_.do_work_ <= 1;
  @ (posedge (wishbone_driver_interface_.work_done_));
  wishbone_driver_interface_.do_work_ <= 0;
  mutex_.put (1);;
endtask
```

By setting `do_work_` to 1, we notify the Verilog of a pending transaction. By waiting for `work_done_` to be 1, we cause the code to wait until the Verilog half of the driver signals that the transaction completed.

The Verilog code is not really interesting, as it, in turn, just calls tasks in a module called `wb_mast`. This module is part of the OPENCORES code. All these files are in the directory `/verification/vip/wishbone`.

This technique of adapting existing Verilog tasks is a good way to leverage working, debugged Verilog code. There is no need to throw the code away, nor any need to rewrite it.

Traffic Generation

Now that we have the chip all configured, we need to send traffic through it. The UART VIP code contained a basic generator, whose interface is shown below:

```
virtual class uart_basic_generator;
  extern function new (string n,
                       uart_configuration c);
  //send one block of words to the uart bfm,
  //hold off sending the block by delay
  extern task send_block (teal::uint32 words,
                          teal::uint32 bit_delay);
  'PURE protected virtual task send_block_
                       (uart_block b);
  endclass
```

The `send_block()` method creates a block of data, with a specific block delay and then calls the connection virtual method `send_block_()`. The data word size is fixed, because the configuration has been randomized previously.

The `send_block_()` is a pure virtual method and is used as the agent connection to the BFM or SFM. The agents are discussed next.

The generator_agent and uart_bfm_agent classes

Now that the generator is generating traffic, we have to connect it to the BFM or SFM. There are as many ways to do this are there are stars in the sky. The authors have chosen to have the connection agents use channels.

```
'include "uart_channel.svh"
class uart_generator_agent extends uart_generator;
   extern function new (string n, uart_configuration c,
                   uart_channel t);
   protected virtual task send_block_ (uart_block b);
      out_.put (b);
   endtask
   local uart_channel out_;
endclass
```

This class does not contain much code. Remember, the purpose of this class within the Truss framework is to enable a connection-policy decision, so the code size is secondary. That said, smaller is better, and this example relies on the channel to do most of the work. It simply puts into a channel the data to be sent.

Let's take a look at the other half, the connection to the BFM or SFM. We'll only show the BFM as the SFM is strikingly similar. Of course, these agents should use a channel as well, because the testbench connects instances of these two classes together.

This class is shown below:

```
'include "uart_channel.svh"
class uart_bfm_agent extends uart_bfm;
  extern function new (string name,
            virtual uart_interface ui,
            uart_configuration c,
            uart_channel_to_be_transmitted,
            uart_channel received_from_wire,
            teal::uint64 clock_frequency);
  virtual task receive_completed_ (uart_word w);
      uart_block current_rx = new (0);
      current_rx.add_word (current_rx_word);
      received_from_wire_.put (current_rx);
  endtask
  local task do_tx_thread ();
    forever begin
      block current_tx;
      to_be_transmitted_.get (current_tx);
      if (current_tx.block_delay_) begin
        pause_(one_bit_ * current_tx.block_delay_);
      end
    end
    for (int i = 0; (i <= current_tx.max_offset ());
          ++i) begin
        send_word(current_tx.words_[i]);
    end
  endtask
endclass
```


There are a few point of interest in the preceding code. Because UART is a bidirectional protocol, there are two channels. One channel is used to connect to the checker agent, and the other channel is used to connect to the generator agent.

Another effect of the UART being a bidirectional protocol is that there are two methods, one to support each channel. One method is the connection technique of overriding a pure virtual method, in this case, `receive_completed_()`.

The other channel-supporting method is `do_tx_thread()`. As you can probably guess, this method runs in a separate thread of execution. This method first delays the appropriate amount. It then takes the block of data words and sends them, one at a time, to the UART BFM.

There is one more point to make before we move on. A chip might have several ways to drive an interface, such as register, FIFO, or DMA. One would probably write corresponding SFMs and SFM agents.

In general, the agents implement a connection policy by overriding the pure virtual method in the base class. In this example, we used a channel policy.

The Checker

Now that we have the transmit side connected, let's take a look at the checking side. We have already done half the work. The agents will place any received data into a channel. We just need to create the checker agent to connect the channel to the checker, as follows:

```
'include "uart_channel.svh"
class checker_agent extends checker;
    extern function new (string name,
            uart_channel expected, uart_channel actual);
    protected virtual task get_expected_
                                (output uart_block b);
      expected_.get (b);
    endtask
    protected virtual task get_actual_
                                (output uart_block b);
```

```
         actual_.get (b);
      endtask
      protected virtual task more_ (output bit value);
        value=expected_.size ();
      endtask
      local uart_channel expected_;
      local uart_channel actual_;
   endclass
```

The checker agent is providing the connection policy for the checker. As we have used a channel for the connection, channels are used here. The expected channel comes from the generator (through a tee, or tap, described in the next section), and the actual channel is the received channel in the BFM.

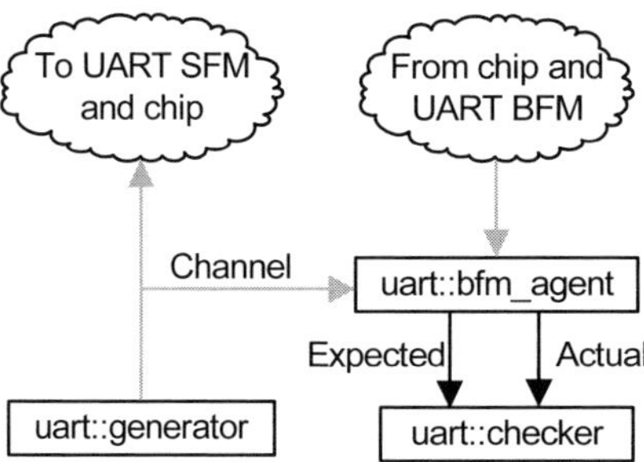

Detailed BFM Agent Connections

The `checker_agent` provides the checker with three things: the expected, the actual, and whether or not there is more checking to do.

Checking the data

Let's take a look at the checker. It's a little complex because the checker must handle the fact that the expected and actual block sizes may be different. This is normal in a real system, for several reasons. One is that a DMA- or FIFO-based receive will "clump up" the received data, depending on how the chip was set up (the specific FIFO interrupt trigger points, DMA block sizes, and so on). Another reason is because a transmission may have to be broken up into segments by the protocol.

Here is the key algorithm in the checker. In the text below, we have removed most of the received data code (because it is identical to the expected code). Also, the code has been simplified just a little, but the essence is still the same. (The actual code is available at www.trusster.com.)

```
task uart_checker::perform_checking_ ();
  uart_block current_tx_block = new (0);
  uart_block current_rx_block = new (0);
  int current_tx = 0;
  int current_rx = 0;
  forever begin
    if (current_tx == current_tx_block.size ()) begin
      get_expected_ (current_tx_block);
      current_tx = 0;
    end
  if (!current_tx.equal (current_rx)) begin
    //... Long error print here!
  end
  //...
  ++current_tx;
  begin
    bit more; more_ (more);
    if ((current_tx == current_tx_block.words_.size ())
        && (!more))
      done_.signal ();
  end
endtask
```

The algorithm compares the data words and relies on dynamic array indexing to move through the block of words. When the array `size()` is reached, a new block of data is pulled from the agent. The algorithm also uses the `uart_word::equal()` to decide how to compare the block elements.

If the agent indicates that there are no more blocks, we signal an event. This event is used by the `wait_for_completion()` code, in a stunning display of software engineering:

```
task uart_checker::wait_for_completion ();
  done_.pause ();
endtask
```

The test, by way of the test component, is waiting for this checker's `wait_for_completion()` to return, signifying that the test is done.

We have made it through the bulk of the code. The next thing we need to talk about is how the objects are built and hooked together. Then we need to talk briefly about the test component and an example test.

Connecting It All Together

The previous sections have discussed the interface components at the various layers of abstraction. Now it's time to put them together. The first place we'll start is the testbench, as it will create the instances and connect them by means of channels. Then we'll take a brief look at the test component, which exercises the ingress or egress flow of traffic. Finally, we'll look at a basic test to send data in both directions.

The testbench

The testbench is responsible for building the protocol components and bringing the chip out of reset. We will not discuss bringing the chip out of reset, as it is pretty much the same as in the tutorial. Building the components, however, is something new.

The components are built in the testbench constructor. We will look at the constructor in stages, as several different things are happening. First, let's look at some naming conventions that will be used in the testbench.

Because the UART protocol is bidirectional, there is a name for the traffic flow in each direction. We will use the industry standard terms of *egress* for traffic originating from the chip and flowing outward, and *ingress* for traffic flowing inward.

Building the channels

Did you notice that the generator, BFM, and checker agents need, among other things, a `uart_channel` in their constructors? Now, here in the testbench, we build a channel, which is shown below:

```
uart_channel program_egress =
    new uart_channel ("program_egress");
uart_channel program_egress_tap =
    new uart_channel ("program egress tap");
program_egress.add_listner (program_egress_tap);
channel program_ingress =
    new uart_channel ("program ingress");
channel protocol_ingress =
    new uart_channel ("protocol ingress");
channel protocol_ingress_tap =
    new uart_channel ("protocol ingress tap");
protocol_ingress.add_listener (protocol_ingress_tap);
channel protocol_egress =
    new uart_channel ("protocol egress");
```

Notice the `add_listener()` call, which connects the `put()` method of one channel to an arbitrary number of other channels. We use it here to give the checker a copy of the generated data.

Building the configuration and interface port

After the channels have been built, there are two more things we need to build before creating the models. They are the configuration and the path to the pins.

Building the configuration is straightforward:

```
uart_configuration = new (n);
```

Building the real (signal) interface to the pins is be done in a few steps. In this example, we first declare the `interface`:

```
interface uart_interface (
    output reg dtr,
    output reg dsr,
    input rx,
    output reg tx,
    output reg cts,
    output reg rts,
    input baud_rate_clock
);
endinterface
```

Next, we declare an instance of this in the `real_interfaces` module:

```
module real_interfaces;
   uart_interface uart_interface_1 (top.DSR, top.DTR,
                   top.TX, top.RX, top.RTS, top.CTS,
                   top.BAUD_RATE_CLOCK);
endmodule
```

Then we define `interfaces_uart`, the class that bundles together *all* the interfaces of the chip. See the file `interfaces_uart.svh`.

Finally, we bundle the interfaces into a class, so that it can be passed into the testbench upon construction:

```
function truss::interfaces_dut build_interfaces ();
  interfaces_uart uart;
  uart = new (real_interfaces.uart_interface_1,
              real_interfaces.wishbone_driver_interface_1,
              real_interfaces.uart_16550_interface_1,
              real_interfaces.top_reset_1);
  return uart;
endfunction
```

Once the testbench recovers this class by means of a downcast, all the interfaces of the chip are available to the verification system.

Building the component-layer objects

Now we are ready to build the components of the protocol, as shown below:[1]

```
begin
uart_bfm_agent ba = new ("uart Protocol",
   uart_dut.uart_interface_1, uart_configuration,
   protocol_ingress, protocol_egress,
   UART_CLOCK_FREQUENCY);
  uart_protocol_bfm = ba;
end
begin
uart_16550_agent sfm   = new (
   "16550 uart",uart_dut.uart_16550_interface_1,
   uart_configuration, program_egress, program_ingress
   UART_CLOCK_FREQUENCY);
  uart_program_sfm = ba;
end
//build and hook up the ingress and egress stimulus
//and scoreboards of the interface
begin
  uart_generator_agent gen_agent = new (
    "egress_generator", uart_configuration,
    program_egress);
  uart_egress_generator = gen_agent;
end
begin
  uart_generator_agent gen_agent = new (
     "ingress_generator", uart_configuration,
     protocol_ingress);
  uart_ingress_generator = gen_agent;
end
begin
  new uart_checker_agent check_agent = new (
    "ingress checker", protocol_ingress_tap,
    program_ingress);
  uart_ingress_checker = check_agent;
end
begin
```

---

[1] In the Layered Approach chapter, we called the *program* side of a chip the registers, and the *protocol* side the wires that follow a standard protocol.

```
    uart_checker_agent check_agent = new (
      "egress checker", program_egress_tap,
      protocol_egress);
    uart_egress_checker = check_agent;
  end
```

The generator and checker are used for both sides of the verification process. This is appropriate, because the generator and checker should have no idea of the connection policy or actual implementation details of the protocol.

Note that the testbench code interface exposes only the base class, not the agents. This allows different connection policies to be invisible to the rest of the verification system. This means that we have to use a trick when we build the derived objects. The trick is to use a local variable of the derived class type and then `new()` that variable. Then we assign the variable to the base class pointer that is our data member.

We are almost done with building all the lower-layer objects. We just need to create the register access objects.

The wishbone objects

Building the wishbone objects is just a matter of building a driver and memory bank, and then mapping the memory bank to an address range, as follows:

```
//in interfaces_uart.svh
  interface wishbone_driver_interface (
    input clock_,
    output reg [31:0] address_,
    output reg [31:0] data_in_,
    input [31:0] data_out_,
    output reg [3:0] select_,
    output reg [1:0] op_code_,
    output reg do_work_,
    input work_done_
    );
  endinterface
//and in interfaces_uart.sv (module real_interfaces)
  wishbone_driver_interface wishbone_driver_interface_1 (
      .clock_ (top.wishbone_driver_verilog.clk),
      .address_ (top.wishbone_driver_verilog.address),
```

```
                    .data_in_  (top.wishbone_driver_verilog.data_in),
                    .data_out_ (top.wishbone_driver_verilog.data_outr),
                    .select_   (top.wishbone_driver_verilog.select),
                    .op_code_  (top.wishbone_driver_verilog.op_code),
                    .do_work_  (top.wishbone_driver_verilog.do_work),
                    .work_done_ (top.wishbone_driver_verilog.work_doner)
                    );
  //and in testbench.svh
  wishbone_driver_ = new ("WB",
                    uart_dut.wishbone_driver_interface_1);
  begin
    wishbone_memory_bank m =
      new("Wishbone",wishbone_driver_);
    teal::add_memory_bank (m);
    teal::add_map ("main bus",
          uart_registers_first, uart_registers_last);
  end
```

The `wishbone_driver` is created and handed to the
`wishbone_memory_bank`, which caches the pointer. Then, the
`wishbone_memory_bank` is added into the Teal memory system. Finally,
this newly added bank is mapped to the first through the last register
address of the UART 16550 interface of our chip.

That's it! From this point in the code and onward, any `teal::write()`
or `teal::read()` to that address range will go through the
`wishbone_memory_bank` and then to the driver.

Whew, that was a lot of code! However, building all the components of
a testbench is a large job. We'll now move up a level, looking at the test
component and then the test.

The test component

Compared to the testbench, the test component is simple. The testbench
pretty much just forwards its dance calls to the appropriate generator,
model, or checker, as follows:

```
class uart_basic_test_component extends
    truss::test_component;
  extern function new (string n, uart_generator g,
        truss::verification_component b, uart_checker c);
  extern virtual virtual task do_randomize();
```

```
//shown in next section
virtual task time_zero_setup ();
    bfm_.time_zero_setup ();
endtask
virtual task out_of_reset(reset r)
    bfm_.out_of_reset (r);
endtask
virtual void write_to_hardware()
    bfm_.write_to_hardware();
endtask
protected virtual void start_components_();
    bfm_.start(); checker_.start();
endtask
protected virtual void do_generate();
  generator_.send_block (words_,block_delay_);
endtask
protected virtual void wait_for_completion_();
  checker_.wait_for_completion();
endtask
rand protected teal::uint32 words_;
local teal::uint32 min_words_;
local teal::uint32 max_words_;
rand protected teal::uint32 block_delay_;
local teal::uint32 min_block_delay_;
local teal::uint32 max_block_delay_;
endclass
```

We won't go over the code above in detail; just take a look and notice that most of the methods are one-line calls to the appropriate interface-layer component.

The last few lines are interesting. They are the random variables that are used by the `do_generate()` method to create random data. These are the variables that will be controlled by the test (as well as by configuration variables).

The uart_basic_test_component::do_randomize() method

The `do_generate()` method is where the test component sends traffic through the interface. It sends only one group of data, but that group length can be any size. The next chapter shows how this method can be called repeatedly.

The `do_generate()` method does only what it is told. The `do_randomize()` method is responsible for choosing the appropriate block length and delay for the block. Why do we separate these two related methods? Because you may want different constraints and distributions for the random parameters. Note the following:

```
task uart_basic_test_component::do_randomize ();
  min_words_ = dictionary_find(name +
        "_min_num_words", 2);
  max_words = dictionary_find(name +
        "_max_num_words", 4);
  min_bit_delay = dictionary_find(name +
        "_min_block_delay", 0);
  max_bit_delay = dictionary_find(name +
        "_max_block_delay", 10);
  'truss_assert (randomize ());
endtask
```

Teal's dictionary is used to see if any high-level code (such as a test) has overridden the parameters. Then, the built-in SystemVerilog function `randomize()` is used to generate the values, subject to the minimum and maximum specified.

That's all there is to the test component. Once a test creates one and follows the standard Truss dance, traffic will be sent and checked through the UART protocol!

Now let's take a look at the test.

The basic data test

The only top-most component that we have not talked about is the test. The test, like the test component, is straightforward. That is as we expect, because the top-most layers should be obvious.

The test is fairly unremarkable. Here is an abbreviated look at its code interface:

```
class block_uart extends truss::test_base;
  extern function new (testbench tb, truss::watchdog wd,
                        string name);
  //... All the usual dance methods, for example...
  virtual task write_to_hardware();
    uart_test_component_egress_.write_to_hardware ();
    uart_test_component_ingress_.write_to_hardware ();
  endtask
  local testbench testbench_;
  local uart_basic_test_component
                        uart_test_component_ingress_;
  local uart_basic_test_component
                        uart_test_component_egress_;
endclass
```

The test builds two test components, one for inbound traffic and one for outbound traffic. For each method, it just calls the same named method on each component.

The authors realize that this can seem tedious, but at least you have all the control. If you need to do some special pre- or postprocessing, it's a simple matter to add it. If you don't want to call all the test components' methods all the time, just leave it out. It there is a specific order you need, or you need some extra communication between the test and the test components, you can just add them.

One alternative, which the authors have used, is to have a global sequencer. This is almost always a mistake, in that it makes the test writer's job harder. Remember the guideline—that "tedious and obvious" is preferable to "less code and hidden."

The interesting part of the test is in the constructor, as shown below:

```
function block_uart::new (testbench tb,
                          truss::watchdog w,
                          string n);
  super.new (n,w);
  uart_test_component_ingress_ = new
        ("uart_ingress", tb.uart_ingress_generator,
        tb.uart_program_bfm, tb.uart_ingress_checker)),
  uart_test_component_egress_  = new
        ("uart_egress", tb.uart_egress_generator,
//add configuration default constraints
  teal::dictionary_put (
    {tb.uart_configuration.name, "_min_baud"}, "4800",
    teal::dictionary_default_only);
  teal::dictionary_put (
    {tb.uart_configuration.name, "_min_data_size"}, "5",
    teal::dictionary_default_only);
  teal::dictionary_put (
    {tb.uart_configuration.name, "_man_data_size"}, "8",
    teal::dictionary_default_only);
  //add generator default constraints
  teal::dictionary_put (
    {tb.uart_egress_generator.name,"_min_word_delay"}, "1"
    teal::dictionary_default_only);
  teal::dictionary_put (
    {tb.uart_egress_generator.name,"_max_word_delay"}, "3"
    teal::dictionary_default_only);
  //...
endfunction
```

This code does two things. First, it creates and wires up the ingress and egress test components. Second, the constructor adds some parameter values to guide the configuration selected and the amount of data to be sent.

That it! We've made it through the first real-world test system!

More Tests

While the test in the example is sufficient for most of the "normal" cases, there are still several things we should do to test the core fully. Besides the additional features of the core, like loopback and FIFO depth triggering, there are a range of error tests to be performed.

For example, one can test parity errors or stop bits, or perhaps the sampling algorithm for the data bits.

There are also the external control pins, such DTR, DSR, and so on, that should be exercised.

All of these tests, which must be written and performed, are beyond the scope of this handbook.

Summary

This chapter ties together the last couple of hundred pages or so. We built a verification system to unit-test a UART.

A configuration convention was covered. Truss does not address chip configuration, because this is chip- and feature-specific. We did show how the Teal dictionary can be used to get and set parameters globally.

An interesting part of configuring the chip was using the Truss register defines with the Teal memory space. This provided a generic register interface that could be mapped to any memory bank. In our case, we adapted a wishbone Verilog model.

The policy of channels was selected to connect the transaction-level classes with the connection-layer ones. We used the Truss pseudotemplated channel.

Checking the data was a little complicated, because the packets to be checked were possibly a different size from when they were generated.

The test component, testbench, and test were described, with an emphasis on the testbench constructor. This was where all the protocol objects were created and the channels connected.

Chip-Level Testing

And will you succeed? Yes indeed, yes indeed!
Ninety-eight and three-quarters percent
guaranteed!

Dr. Seuss

Testing at the block level is common. Testing at the chip level—the highest system level we care about—is necessary. As you probably know, it's the system-level interactions among the various blocks that must be tested. These system-level interactions are the focus of this chapter.

This chapter presents three main concepts:

- The chip now has four UART interfaces.

- We develop three tests, showing a progression from exercising all of the protocols to demonstrating a generic test with irritators.

- We can adapt the original block-level test to be used at the system level.

Overview

This chapter highlights Truss irritators. We'll adapt the UART block-level testbench to a system-level testbench that has four UARTs. One of these UARTs will be randomly chosen to be the focus of the test, while the other three will serve as background traffic irritators. While this chapter uses UARTs for the irritators, the idea is generic.

Theory of Operation

This verification system builds upon the block-level UART system. We will adapt the components developed in the last chapter, and add a few new tests. These tests will show how irritators are used.

Here are the main components involved in the simulation:

Quad UART Example: Objects and Connections

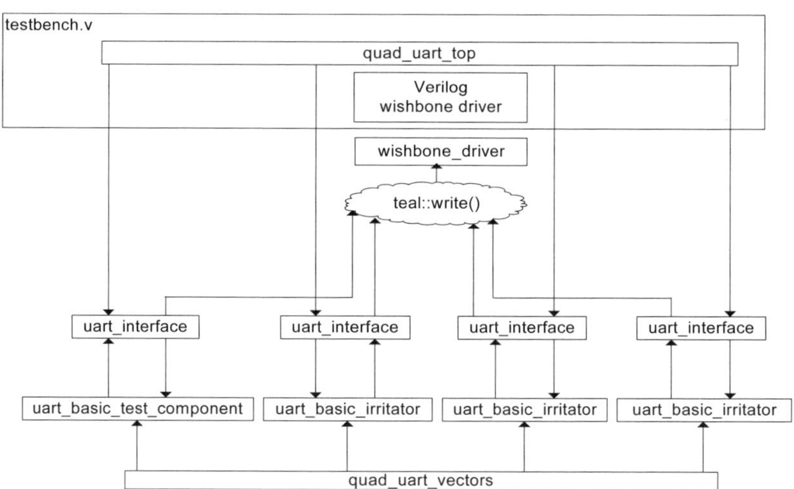

One difference from the block-level system is that the testbench now does not directly build each of the four UART interface's objects (such as the generator, checkers, agents, and so on). This is left to the `uart_group` class, which is called by the testbench. Another difference is that three of the interfaces are exercised by irritator objects.

As with all good projects, this project steals[1] code (in our case, from the block-level UART code). The UART BFM, generators, and checkers are reused without modification. We modify the UART 16550 SFM by adding an integer ID, which is used to form the specific UART address.

The UART test component is also reused directly. In addition, we inherit from the test component to create an irritator class. So, we added only a few lines of code to the block-level test.

We will develop four tests. One is the previous chapter's block-level test, with modifications to select one of the four UARTs. The other three tests exercise all four UARTs at once and show the test development progression from a simple group of test components to a single test component and a list of irritators. Because the irritators are a Truss common base class, this final test can be used as model for random tests in the Truss environment.

Verification environment

Looking at the verification environment, we see that it is very similar to that for the single UART. We still have the wishbone driver to access the registers, this time mapped to a larger region. We still have all the main players from the block-level UART.

Looking at the HDL side, the testbench environment is fairly straightforward. The only different module, `quad_uart_top`, instantiates four UARTs and maps their write enable and read select according to the upper address bits. The testbench environment is shown below.

[1.] Uhhh, adapts.

Quad UART Example: HDL Connections

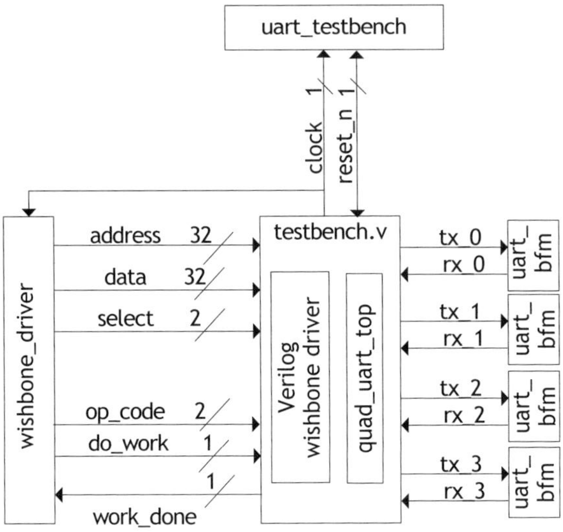

Running the UART Example

There are four tests in the example. Running an example is simple in Truss. To run a test, type one of the following:

```
$TRUSS_HOME/bin/truss --test block_uart --config block
$TRUSS_HOME/bin/truss --test quad_test_components
$TRUSS_HOME/bin/truss --test quad_uart_irritators
$TRUSS_HOME/bin/truss --test quad_uart_vectors
```

The quad_uart_test_components Test

The `quad_uart_test_components` test is the first test for the chip that the authors wrote. We were just making sure that all the UARTs could be addressed. This test chooses four different random configurations and sends a random-length block (with random block delays) to each ingress and egress channel. Here is the constructor part of the test:

```
function quad_uart_test_components::new (
              testbench tb, truss::watchdog w, string n);
  super.new (n, w);
  testbench_ = tb;
  'truss_assert (number_of_uarts >= 2);
  for (teal::uint32 i(0); i < number_of_uarts; ++i) begin
    string id = $psprintf ("%0d", i);
    uart_test_component_ingress_[i] =
      new ({"uart_test_component_ingress ", id},
          tb_.uart_group[i].uart_ingress_generator,
          tb_.uart_group[i].uart_program_sfm,
          tb_.uart_group[i].uart_ingress_checker);
    standard_generator(
      tb_.uart_group[i].uart_ingress_generator.name);
    uart_test_component_ingress_[i].do_randomize ();
    uart_test_component_egress_[i] =
      new ({"uart_test_component_egress ", id},
          tb_.uart_group[i].uart_egress_generator,
          tb_.uart_group[i].uart_protocol_bfm,
          tb_.uart_group[i].uart_egress_checker);
    do_generator (
      tb_.uart_group[i].uart_egress_generator.name);
    uart_test_component_egress_[i].do_randomize ();
    do_configuration (
      tb_.uart_group[i].uart_configuration.name);
  end
endfunction
```

This example scales up nicely from the block-level tests we have seen.
The rest of the test's methods are fairly boilerplate, and do not need any
special attention.

The quad_uart_irritators Test

The `quad_uart_irritators` test is the second test for the chip that the authors wrote. In this test we have randomly selected one UART as the focus of the test (that is, by using a `test_component`), and we have an array of three UART irritators. What are these `uart_irritators`? Let's take a look at the class.

UART irritator class

Remember that a Truss irritator is a way to adapt a test component to run as background traffic. The test component's `run_traffic()` method will be called repeatedly until the test decides its time to stop. Here is the interface for the UART test irritator:

```
class uart_basic_irritator extends truss::irritator;
  local uart_basic_test_component basic_test_component_;
  function new (string n,
               uart_generator g,
               truss::verification_component b,
               uart_checker c):
   super.new(n);
   uart_basic_test_component_ = new (n,g,b,c);
  endfunction
  virtual task out_of_reset()
    basic_test_component_.out_of_reset();
  endtask
endclass
```

The `uart_basic_irritator` class is a `truss::irritator` and owns a `basic_test_component`. This makes sense, as the code interface will follow the Truss irritator model and the implementation will use the `uart_basic_test_component`, which is unchanged from the block-level test. This is a major advantage of Truss: the same test component can be reused from the block level to the system level. Also, because a testbench is not mentioned anywhere in the test component, the test component can be moved to different projects easily. This is not accidental, but rather is a direct result of the layered approach talked about earlier in this handbook.

Let's take a look at the rest of the methods.

```
virtual function void report (string prefix);
  basic_test_component_.report (prefix);
endfunction
virtual task time_zero_setup ();
  basic_test_component_.time_zero_setup ();
endtask
virtual task out_of_reset (reset r);
  basic_test_component_.out_of_reset (r);
endtask
virtual task write_to_hardware ();
  basic_test_component_.write_to_hardware ();
endtask
virtual task wait_for_completion ();
  basic_test_component_.wait_for_completion ();
endtask
virtual protected function void do_randomize ();
  basic_test_component_.do_randomize ();
endfunction
virtual protected task wait_for_completion_ ();
  basic_test_component_.wait_for_completion_ ();
endtask
virtual protected task start_components_ ();
  basic_test_component_.start_components_ ();
endtask
virtual protected task do_generate ();
  basic_test_component_.do_generat2();
endtask
virtual task inter_generate_gap ();
  checker_.wait_actual_check();
endtask
```

Notice that all the methods (save one) just call their
basic_test_component method. This is a standard form, tedious
indeed, but it gives the coder the ability to add special code if needed.
Again, this is the tedious but obvious guideline coming into play.

There is one method, inter_generate_gap(), that is not just calling
the test_component's method. This is because this method is specific
to an irritator. In our case, we know that the checker is derived from the
Truss checker, and so has a method to wait until expected or actual data
are checked. This is an appropriate throttling method for our irritator.

As coded here, this method pauses the generation until a data packet is checked. We could have done fancier things, such as have an initial number of packets in play, or change the delay depending on the actual data bytes generated.

That's it for the irritator! In less than two dozen lines code, we have added the ability to use any UART interface as background traffic. Furthermore, the interface is that of a generic irritator, able to be plugged into any test that has a list of `truss::irritators`. This is shown in `quad_uart_irritators`, the first test to use irritators. Let's take a look at the test.

The test

Of course, we also have a test component that is the focus of the test. This test is a little confusing, in that we use a UART for both the test component and the irritators. Nevertheless, this is what we have to test. Here are the interesting parts of the test's code interface:

```
parameter uint32 irritator_count = number_of_uarts - 1;
class quad_uart_irritators extends truss::test_base;
  extern function new (testbench tb, watchdog w,
                       string name);
  //...
    local testbench testbench_;
    //The focus of the test
    local uart_basic_test_component uart_ingress_;
    local uart_basic_test_component uart_egress_;
    //The background traffic components
    uart_basic_irritator
        uart_irritator_ingress_[irritator_count];
    uart_basic_irritator
        uart_irritator_egress_[irritator_count];
endclass
```

This test uses a fixed array of irritators. In this test they are explicitly called out as `uart_basic_irritator`. This use of a specific irritator type will be made more generic in the next test.

Let's take a look at the implementation of the constructor:

```
function quad_uart_irritators::new(testbench tb,
                              truss::watchdog w, string n);
  super.new (n, w);
  testbench_ = tb;
  'truss_assert(number_of_uarts >= 2);
endfunction
```

Where did all the code to initialize the test components go? That code is moved into the `do_randomize()` method, because now the test has some random behavior. In this case, randomization determines which UART interface to pick for the test_component. Here is the `do_randomize()` method:

```
task quad_uart_irritators::do_randomize();
//First, for the  main point of the test...
  min_index =
    dictionary_find ({name_,"_min_uart_index"}, 0);
  max_index =
    dictionary_find ({name_,"_max_uart_index"}, 0);
'truss_assert (randomize ());
  log_.info($psprintf ("Selected uart %0d", uart_index_));
  begin
    int i = uart_index_;
    string id = $psprintf ("%0d", i);
    uart_test_component_ingress_[i] =
      new ({"uart_test_component_ingress ", id},
           tb_.uart_group[i].uart_ingress_generator,
           tb_.uart_group[i].uart_program_sfm,
           tb_.uart_group[i].uart_ingress_checker);
    //...
    standard_generator (
    testbench_.uart_group[i].uart_ingress_generator.name);
  uart_test_component_ingress_[i].do_randomize ();
  uart_test_component_egress_[i] =
    new  ({"uart_test_component_egress ", id},
      tb_.uart_group[i].uart_egress_generator,
      tb_.uart_group[i].uart_protocol_bfm,
      tb_.uart_group[i].uart_egress_checker);
  //...
  do_generator (
    testbench_.uart_group[i].uart_egress_generator.name);
```

```
      uart_test_component_egress_[i].do_randomize ();
  do_configuration (
    testbench_.uart_group[i].uart_configuration.name);
  end
  begin
    //now for the irritators...
    teal::uint32 count = 0;
    for (teal::uint32 i(0); i < number_of_uarts; ++i)
      begin
        string id = $psprintf ("%0d", i);
        uart_group if = tb_.uart_interface[i];
        'truss_assert (count < irritator_count);
      if (i != uart_index_)
        begin
          uart_irritator_ingress_[count] =
            new ({"irritator_ingress ",id},
              if.uart_ingress_generator,
              if.uart_program_sfm,
              if.uart_ingress_checker);
          uart_irritator_egress_[count]   =
            new ({"irritator_egress ",id},
              if.uart_egress_generator,
              if.uart_protocol_bfm,
              if.uart_egress_checker);
          do_generator (if.uart_egress_generator.name);
          do_generator (if.uart_ingress_generator.name);
          count++;
        end
      end
    end
  endfunction
```

Okay, that code is a bit long—but it is straightforward. First, a UART protocol is chosen to be the focus of the test. Then it is built and randomized. After that, the rest of the UART interfaces are packed into test irritators and randomized.

Now that all the components have been built, let's look at a typical test method:

```
task quad_uart_irritators::start ();
  uart_test_component_ingress_.start ();
  uart_test_component_egress_.start ();
  for (teal::uint32 i = 0; i < irritator_count; ++i)
  begin
    uart_irritator_ingress_[i].start ();
    uart_irritator_egress_[i].start ();
  end
endtask
```

All the methods of the test follow this form. They first perform the action on the `test_component`, and then on the irritators. The `wait_for_completion()` is similar:

```
task quad_uart_irritators::wait_for_completion( );
  uart_test_component_ingress_.wait_for_completion ();
  uart_test_component_egress_.wait_for_completion ();
  for (teal::uint32 i = 0; i < irritator_count; ++i)
  begin
    uart_irritator_ingress_[i].stop_generation ();
    uart_irritator_egress_[i].stop_generation ();
  end
  for (teal::uint32 i = 0; i < irritator_count; ++i)
  begin
    uart_irritator_ingress_[i].wait_for_completion ();
    uart_irritator_egress_[i].wait_for_completion ();
  end
endtask
```

Notice that `wait_for_completion()` first waits for the focus of the test to complete. Then it tells the irritators to stop, and then waits for the irritators to complete.

Remember, that after this `wait_for_completion()` returns, the test is done.

The quad_uart_vectors Test

The `quad_uart_vectors` test is the logical evolution of the previous test. We get more fancy. Instead of a fixed array, we use a dynamic array. Here are the relevant parts of the test header file:

```
class quad_uart_vectors extends truss::test_base;
  extern function new (testbench tb, truss::watchdog w,
                       string name);
//standard Truss methods not shown
//as before:
  local testbench testbench_;
  local uart_basic_test_component uart_ingress_;
  local uart_basic_test_component uart_egress_;
//new stuff!
  truss::irritator irritators_ [$];
endclass
```

The `do_randomize()` method is very similar to that used in the `quad_uart_test_components` test, with a small difference. Here we build irritators, not test components.

```
uart_basic_irritator bi =
    new ({"uart_irritator_ingress ", id},
        if.uart_ingress_generator,
        if.uart_program_sfm, if.uart_ingress_checker);
irritators_.push_back (bi);
```

The methods all follow a standard form, but for those not familiar with the macros, they can look unnatural. Here is one example method:

```
'define for_each(data, method)\
    for (integer i = 0; i < data.size (); i++) \
      begin \
       data[i].method (); \
      end
task quad_uart_vectors::time_zero_setup ();
  uart_test_component_egress_.time_zero_setup ();
  'for_each (irritators_, time_zero_setup);
endtask
```

The `'define` allows us to operate on an entire array with a small amount of code.

The methods are just a little more complicated for the Truss methods that have a parameter:

```
'define for_each_1(data, method, param)\
   for (integer i = 0; i < data.size (); i++) \
     begin \
       data[i].method (param); \
   end
function void quad_uart_vectors::report (string p);
  uart_test_component_ingress_.report(p);
  uart_test_component_egress_.report(p);
  'for_each (irritators_, report, p);
endfunction
```

This last form of the test contains the fewest lines and also uses a macro. It's up to you to decide whether this is appropriate for your system.

The block_uart Test

The `block_uart` test is just a rework of the block-level test. The only changes were to use the testbench's `uart_interface` objects, and to select a protocol to exercise.

Summary

This chapter took a look at a system-level verification system. We adapted the components from the block-level test.

The first test just re-used the `test_component` from the block-level test on all four UART protocols.

The next test brought in the concept of irritators, background traffic for the main test.

The system-level test `quad_uart_vectors` was used to show how dynamic arrays and macros can be harnessed to make small, efficient, standard-form code.

In general, this chapter showed that many block-level components could be adapted without modification to the chip-level testbench. We did, however, need to modify the `uart_16550_sfm` to handle a specific address range.

Things to Remember

"There goes my tail again." —Eeyore
Paraphrased from Winnie-the-Pooh, by A.A. Milne

An ending is, by definition, a new beginning. This, the last chapter, provides a good opportunity to review some of the handbook's main points. The authors sincerely hope that this is also a beginning for you to benefit from using some of the techniques presented in the preceding chapters.

This chapter is the 30,000-foot view of what we have covered. A wise, experienced manager once told the authors, "If you want your team to remember something, tell them at most three things." We take that advice—sort of—and present the three most important ideas of each part in the book.

We hope that this handbook, and its accompanying code, was and will continue to be useful. In the end, however, it is your job to verify the chip.

Part I: Use SystemVerilog and Layers!

In the first part of the book we introduced verification, SystemVerilog, object-oriented programming, and what a layered verification looked like. Here are the important points:

- SystemVerilog is a good language for verification.

- Use OOP techniques for verification, but not to excess.

- Layering is the main technique for a verification system.

The verification world is a bit enamored with OOP. We are probably in the early stages of settling down and using it, or not, where appropriate. By using OOP techniques we can communicate our architectural intent clearly.

The concept of layering, formally described as abstraction, roles, and responsibilities, is perhaps the single most important technique we can use. We presented terms for layers that we later implemented as classes and conventions.

Part II: An Open-Source Approach

In this part of the handbook we presented some code that has proved useful to us and those at other companies. That code may not have everything you want, but it should be flexible enough for you to adapt it to your needs. We noted specifically the following:

- Teal is a set of useful classes and features for verification. These are the building blocks of functional verification.

- Truss provides a flexible, yet well-defined, application framework for verification.

- A simple, but complete, example can be useful.

This part of the handbook is what most books lack. The authors take all the theory and lessons learned and show you how they have built verification systems. Make no mistake, Truss is a verification methodology. Teal is a bit more open, but any implementation of a programming concept contains the prejudices and biases of the implementors.

The point of the example is to show how these implementations, Teal and Truss, can be useful.

Part III: OOP—Best Practices

In this part of the handbook we took a long look at OOP. We talked about how to "think OOP" and how to "code OOP." Here are the three main points of this section:

- OOP is a powerful tool for managing complexity and creating adaptable code.

- There are lots of techniques, and most of them involve balancing trade-offs.

- The code should make minimal assumptions, and make those assumptions as obvious as possible.

As the complexity of the chips increased, so did our verification systems. OOP can be used to increase the communication among engineers. Basically, this means creating code that others can reason about.

The hundred pages or so of the middle part presented lots of lessons learned. There were techniques, guidelines, and horror stories. There were no absolute right or wrong answers. You and your team must decide what is appropriate.

If a bit of code has a well-defined purpose as well as obvious dependencies, it stands a good chance of being reasoned about and eventually understood. The objective is to minimize the assumptions about the code, while still doing something worthwhile.

Part IV: Examples—Copy and Adapt!

We could have left the book with only three parts—but one example and some code snippets are usually not enough to help you understand a set of techniques or some new code. Consequently, we wrote some more examples. The following summarizes the main points of these examples:

- You can create portable verification IP that other projects can use.

- Separating the chip-specific parts from the protocol-generic parts shows users what they have to modify for their project.

- The testbench and test components can become large, but they are still "reasonable."

This section of the handbook presented more examples of chips and their verification systems, all the way to a final example that used all the previous examples. We would not be surprised if the code has mistakes and can be made even more clear. Yes, we'll probably even get some complaints, but we know of no better way for you to learn the ideas and techniques talked about in this handbook than to see working, completed examples.

Conclusion to the Conclusion

The authors have tried to make a handbook that is useful. We've combined verification and OOP, and described techniques that have proved useful.

We did not separate the verification techniques from the language used to express them. To do that would have made the book easier to write. However, you would have been reading just a book, not a handbook. You should be able to find in these pages—and on the code freely available at www.trusster.com—enough examples that are sufficiently close to what you want to do. Cut, copy, and paste away!

Please contact us at www.trusster.com. On this site you can also find up-to-the-minute information about Teal and Truss, as well as discussion boards where users share knowledge and ideas. It's a good place to start for any Teal or Truss questions.

It's also where we will post errors found in this handbook. We invite your comments and suggestions.

Please stay in touch with us, at www.trusster.com.

" . . . and now for something completely different.[1]"

---

[1] From Monty Python's Flying Circus, episode 26, December 1971.

Index

Symbols

Hardware verification with C++